THE GENE
MAKEOVER

THE 21ST CENTURY ANTI-AGING BREAKTHROUGH

VINCENT C. GIAMPAPA, M.D., F.A.C.S.

FREDERICK F. BUECHEL, M.D., F.A.A.O.S., F.A.C.S.

AND OHAN KARATOPRAK, M.D.

Basic Health
PUBLICATIONS, INC.

The information contained in this book is based upon the research and personal and professional experiences of the authors. It is not intended as a substitute for consulting with your physician or other healthcare provider. Any attempt to diagnose and treat an illness should be done under the direction of a healthcare professional.

The publisher does not advocate the use of any particular healthcare protocol but believes the information in this book should be available to the public. The publisher and authors are not responsible for any adverse effects or consequences resulting from the use of the suggestions, preparations, or procedures discussed in this book. Should the reader have any questions concerning the appropriateness of any procedures or preparation mentioned, the authors and the publisher strongly suggest consulting a professional healthcare advisor.

Basic Health Publications, Inc.

28812 Top of the World Drive • Laguna Beach, CA 92651

949-715-7327 • www.basichealthpub.com

Library of Congress Cataloging-in-Publication Data

Giampapa, Vincent C.
   The gene makeover : the 21st century anti-aging breakthrough / Vincent C. Giampapa, Frederick F. Buechel, and Ohan Karatoprak.
      p. cm.
   Includes bibliographical references and index.
   ISBN 978-1-59120-198-4
   1. Longevity. 2. Genetic regulation. 3. Dietary supplements. I. Buechel, Fred.
II. Karatoprak, Ohan. III. Title.

   RA776.75.G537   2007
   613.2—dc22
                                                2007031928

The artwork in this book is used with permission of the following:
Am. J. Clin. Nutr. (2006;83:436S–442S), *American Journal of Clinical Nutrition* (adapted by
   Dr. Giampapa): Figure 6.1
BioMarker, Inc.: Figures 1.1, 1.6
Giampapa, Vincent, M.D., F.A.C.S. *The Basic Principles and Practice of Anti-Aging Medicine* &
   Age Management for the Aesthetic Physician and Surgeon. © 2003: Figures A.1, A.2, A.3, A.4
The Giampapa Institute for Anti-Aging Medicine: Figures 1.2, 1.5, 2.1, 2.2, 2.3, 2.4, 2.6, 3.1, 4.1, 4.2,
   4.3, 4.4, 5.1, 6.2, 6.3, inset in 6, B.1, B.2, B.3, B.4.
J. Craig Venter Institute: Figure 1.3
National Human Genome Research Institute: Figures 2.7, 2.8, 2.9
Suracell Inc.: Figures 1.4, 2.5, 7.1, 7.2, 7.3, 7.4

Editor: Karen Anspach • Typesetter/Book design: Gary A. Rosenberg
Cover design: Mike Stromberg

Printed in the United States of America

10   9   8   7   6   5   4   3   2   1

Contents

Prologue

As a child, I wondered when would be the most exciting time to be alive. The Italian Renaissance, I thought, with Leonardo da Vinci, Michelangelo, and Galileo. These brilliant minds were filled with new ideas about human potential. They offered new ways of looking at the world and their environment.

Over a decade ago, a handful of visionary scientists met in Chicago. This group believed there was enough valuable scientific information available to begin to control the aging process, at least to a degree. They also believed that humans could live longer, healthier lives. This was revolutionary thinking that ran contrary to the thinking of the time. With this ambitious purpose in mind, the American Academy of Anti-Aging Medicine was founded. This organization is now the largest scientific group in the world focused on bringing the latest, most credible information about living longer and healthier lives to the public.

As one of the founders of this organization and as the first president of the American Board of Anti-Aging Medicine, I felt then—and even more now—that we possess the vital information to improve the aging process for every individual who desires to use it.

The road to accomplishing this was difficult and challenging. Trained as a plastic surgeon, I was always fascinated by what drove the process of aging. As I watched a beautiful face turn wrinkled and old, I wondered what was occurring at the *deepest* level—the level of the cell. As a plastic surgeon, I am intimately familiar with all the latest surgical procedures and technology used clinically to reverse the signs of aging. Yet, what were we actually doing? We were not really focusing on the causes of this process. Rather, we were merely *illusionists,* covering up and hiding the signs or symptoms of aging. We were not reaching to the fundamental causes of the process itself.

Over the last fifteen years, with the new research and amazing scientific information obtained from the Human Genome Project, we have begun to understand the process of aging at its core. Using that information, we know it is possible for everybody to age more optimally. It is essential that this new information about aging optimally be utilized now more than ever, because as individuals we can and must maintain control of our own health and longevity.

Today, we understand that the secret of aging is contained in our DNA. In each one of the 100 trillion cells that make up the human body there is six feet of DNA. The secret to aging well resides in these billions of miles of DNA that define us as human beings. So the best approach to control aging seems to be by promoting health *one cell at a time,* for it is here that the secret lies. *If we can control the aging process of one cell, then it is possible to control it in the 100 trillion cells making up the human body.*

But first many questions needed to be answered. Since each of us is unique and has different DNA, how would this actually work? Were there any examples of groups of humans throughout history who actually lived longer with better health than most of us? As my colleagues and I explored this process, the answers began to appear. We also began to wonder what was so different about the people we found that met these criteria, and what was different about their DNA. How did their unique DNA interact with their lifestyle and their diet? Was there really any scientific proof that we could make animals and people live longer? Was there an anti-aging drug, pill, or supplement that really worked? All these questions, as well as the answers, slowly began revealing themselves year after year.

And most important, how could we take this information and give it to the entire world so we all might live longer and more healthful lives? That information is the critical story in *The Gene Makeover.* Once our genetic inheritance is known, we can take steps to impact the expression

of our genes so they work more efficiently. That's right. We can genetically enhance ourselves even at this very moment!

As you read this book, you will learn how you can accomplish this
amazing feat. You can find out which of your inherited groups of genes are
actually working more or less efficiently to control the mysterious, underlying process of aging. And with my Personal Genetic Health approach,
you will learn how this may affect your health span—the amount of time
you live enjoyably with a high quality of health—and how to take action
to live the longest, healthiest life possible.

For the very first time in history, human beings have been empowered
with the information about how to live a longer, healthier life coupled
with their own control of the aging process. The only thing you need to
do to accomplish them yourself is to put the information in this book
to use.

Knowing all this makes me think that now is the most exciting time in
history to be alive. In fact, centuries from now humans will look upon this
period as *the true renaissance of medicine,* when humans became masters
of their fate rather than victims of their genes for the first time.

I invite you to join me on this wonderful new journey to enhanced
health and longer life.

Yours for health and longevity,
Vincent Giampapa, M.D., F.A.C.S.

1

Proof We Can Slow Aging Now: Why It Doesn't Have to Be Like It Was for Your Parents!

So many people spend their health gaining wealth,
and then have to spend their wealth to regain their health.

—A.J. Reb Materi

ooking young and staying young are part of the twenty-first century lifestyle. That has created two mega industries: the vitamin industry, which has grown to yield more than $30 billion a year, and the anti-aging industry, which produces more than $100 billion a year. Both of these mega-health industries are growing more rapidly as more people are getting older.

But while more and more people continue to use health products, confusion still reigns in both areas. Based on our medical experience, we find that most people are still looking for definitive answers to such health and nutrition related questions as:

- Why do I need to take vitamin supplements if I eat well?

- Can I really extend my health span or life span by eating right, exercising, and taking supplements?

- What should I take, how much should I take, and when should I take these vitamins and other dietary supplements?

- Are too many supplements bad for me?

- What is anti-aging really? Is it hormone replacement? Is it growth hormone? Is it lifestyle changes like incorporating a new diet, exercise, or other things into my daily routine?

- Can I reduce my risk of age-related diseases like cancer, cardiovascular diseases, diabetes, Alzheimer's, osteoporosis, depression, anxiety, and arthritis?

- If I already have a disease, can I make easy and realistic lifestyle changes to be healthier and stop or even reverse its progression?

Our purpose in writing this book is to answer these and other questions—to end the confusion once and for all. Now it's possible for every person to stay healthy and productive as long as possible. The information needed to accomplish this personal health goal is now available from a series of recent medical breakthroughs about extending life span and from our new understanding of human genetics. These recent scientific developments prove that age management is really about using the genes we've inherited to their optimal potential. The secrets to a longer, healthier, and better quality of life are hidden inside each of our cells—in our genes.

COULD THE FOUNTAIN OF YOUTH REALLY BE RELATED TO WHAT YOU DRINK AND EAT?

History is filled with tales of people searching for the proverbial fountain of youth or, in modern times, the magic pill. Most people in the United States know about how the famed search of Ponce de Leon ended in the discovery of Florida, but no fountain of youth. It wasn't until recently that explorers of a different kind began making discoveries leading to extending life span, and they have found that you don't have to travel around the world like Ponce de Leon and other longevity seekers did in the past. An important part of the longevity secret can be found right in your kitchen, with the very foods you eat every day. Yes, the most compelling research has concluded that one of the secrets to longevity rests in the very hand that feeds you.

The Proof We Can Slow Aging and Increase Our Health Is Now Here: Calorie Restriction

For several decades studies were conducted on rats, chimps, other animals, and even microbes (bacteria) with the sole purpose of figuring out how to make them live longer. Some of the early headlines that found their way from the medical journals into the general press touted how these early discoveries resulted in extending the life of yeast, earthworms, and rats. However, the secret of longevity these researchers discovered was not a magic potion handed down from the Garden of Eden or a new drug. The secret to longevity based on these early research studies was related to eating less food.

Calorie Restriction Makes Animals Live Longer and Healthier

Contemporary researchers give credit to an early, pioneering study reported in a 1935 issue of the *Journal of Nutrition* by Dr. C. M. McCay and Dr. L. A. Maynard about calorie restriction and increasing life span in animals. Back then this was indeed a revolutionary idea that captured the imagination of people around the world. They reasoned that if the life span of animals could be increased, then increasing the life span of humans was within the grasp of science. (See Figure 1.1.)

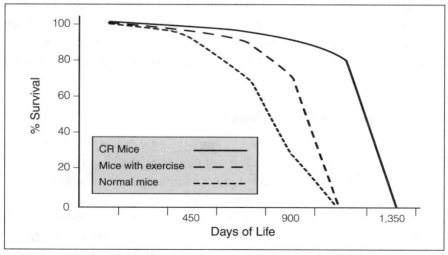

Figure 1.1. Study shows 30 to 40 percent longer, healthier lives with calorie restriction.

The Oxidative Stress Factor

Researchers everywhere began experiments using all sorts of animals: from single-celled microbes to slightly more complicated invertebrate worms, from fruit flies to fish, and from rats to monkeys, eventually making their way to humans. Once the general idea was established that calorie restriction could increase life span, the next step for the researchers was to determine why. What was it about restricting calories that made animals live longer? From these decades of research some important insights were discovered about slowing down aging, reducing diseases, and increasing life span. This early research also determined that animals need the right balance of reduced caloric intake and essential nutrients to experience optimal longevity.

One of the longevity factors revealed from the animal studies was a reduction in what scientists call oxidative stress. Oxidative stress is part of living. It occurs all the time. A major source of oxidative stress comes from the compounds released when we digest food. So it follows that when fewer calories are eaten, internal oxidative stress is automatically reduced.

But for humans, reducing our naturally occurring internal oxidative stress is just part of the challenge. In these times extra oxidative stress also comes from chemicals outside our bodies, so we have too much oxidative stress in our lives, and our protective systems are overwhelmed. This led scientists to the idea that ingesting larger amounts of nutrients with what's called "antioxidant" activity would be a way to help reduce oxidative stress. The most recent studies do indeed support this contention, as do our own results of working directly with people, which we will elaborate upon in Chapter 6.

More Longevity Benefits Revealed

As this body of research evolved, scientists began to better understand how basic biology common to all animals is related to longevity and health. These vital functions eventually became referred to as biomarkers. Some of the biomarkers associated with calorie restriction and increased life span in animals include: reduction in body temperature; reduced energy production; decreased levels of insulin; improved insulin sensitivity and functioning; less body fat; lower blood pressure; reduction in hardening of the arteries; less clogging of arteries; decreased inflammation; and favorable hormone balance, including DHEA (dehydroepiandrosterone), which is considered to be a master-control and youth-promoting hormone produced in some animals and in humans. As research progressed, it was discovered that these biomarkers can be influenced by lifestyle, and that major improvements in health can be experienced by taking the right supplements. This will be reviewed in detail in Chapter 6.

Controlling Gene Expression

The most recent animal studies report how calorie restriction can affect gene expression. Your genes are stored in each cell on strands of DNA (deoxyribonucleic acid), called chromosomes. Twenty-four hours a day, the genes on your chromosomes are being turned on and off to make the substances you need to function and stay alive. When your gene expression is functioning properly, you are healthy. When it is not functioning properly, your health is adversely affected, your life span is reduced, and the diseases of aging begin to appear.

While we will be discussing this information in more detail in subsequent chapters, it is important to realize from the start that aging is associated with changes in gene expression throughout the body. These changes are associated with the decline in brain, heart, digestive system, and muscle functions that are related to getting older.

The good news is that life span animal studies show calorie restriction seems to be able to slow down the normal decline in body function. This is thought to occur by increasing protein synthesis and by decreasing damage to cells and molecules in the body, including, most important, DNA. This means less damage to cells and tissues leading to better cell growth, function, and repair—the very things we need to live longer and healthier lives.

RESEARCH INTO HUMAN LIFE SPAN

While nonhuman calorie restriction research was underway, a special group of researchers called epidemiologists started looking around the world for groups of people who have been practicing calorie restriction as a natural way of life. Researchers found what they were looking for in the 1990s on a tiny island just south of mainland Japan. People living on Okinawa were healthier and a higher number lived to be 100 compared to populations studied elsewhere. For example, compared to Americans, Okinawan seniors:

- Are 75 percent more likely to retain cognitive ability
- Get 80 percent fewer breast and prostate cancers
- Get 50 percent fewer ovarian and colon cancers
- Have 50 percent fewer hip fractures
- Have 80 percent fewer heart attacks
- Have an average life expectancy of eighty-seven years, compared to an American male average life expectancy of seventy-eight years
- Have the largest number of centenarians

What Enables More People on Okinawa to Live Longer and Healthier?

While genetics is always a factor in health, researchers determined that the Okinawan diet and lifestyle play the major role in this spectacularly healthy aging phenomenon. Not only is the typical Okinawan diet lower in calories when compared to the average American diet, it is also healthier. The Okinawan diet is low in fat, animal foods, and junk foods while it's high in fruits and vegetables, fiber, omega-3 fatty acids, whole foods, and bioflavonoids. In fact, the Okinawan biomarkers of longevity and health were similar to biomarkers in animal studies on calorie restriction.

For the first time, there was evidence of a correlation between the insights from calorie-restricted life span research in animals and similar life span mechanisms in humans. But more research was needed to unlock the much-sought-after secret of biological aging and how to increase life span in humans.

More Promising Research from the East

Additional research conducted in Japan supported the association between energy intake and life span. When researchers followed a group of 1,915 healthy men for thirty-six years who were nonsmokers aged forty-five to sixty-eight, they found what they were hoping for. In 2004 the team of researchers headed by Dr. Bradley Willcox observed a trend toward increased life span in the group of men ingesting about 15 percent fewer calories per day below the mean intake. This group of reduced-caloric-intake men also had reduced incidence of diseases.

Thus far, based on epidemiological research, it seemed that the longevity and disease-prevention benefits of calorie restriction observed in animal species might also hold true for people. But to be sure, more proof was needed. Surprisingly, another round of proof came almost by accident from an ecosystem research project called Biosphere 2.

Biosphere 2 Provides More Longevity Evidence

Remember Biosphere 2? In that experiment in the 1990s people were confined for two years in an enclosed but self-sustaining ecosystem. The inhabitants primarily ate a nutrient-dense, vitamin-supplemented vegetarian diet that was also calorie controlled. The original intent was for the inhabitants to eat their nutrient-dense diet at the rate of consumption equal to the daily calories they expended—just what they needed for weight maintenance and daily activities.

However, food production started to lag during the experiment, and for several months the inhabitants had to sustain themselves on a low-calorie diet between 1,700 to 2,100 calories per day. This level of daily caloric intake represented a calorie-restriction diet in which the caloric intake was less than the calories normally used each day. Unexpectedly, the Biosphere 2 experiment turned out to be the first human calorie-restriction experiment for part of two years.

The Biosphere 2 inhabitants were routinely examined by medical professionals, so there was a good set of data about the condition of their health. Researchers were able to analyze the data to determine the effects of calorie restriction. Did the prolonged calorie restriction of the Biosphere 2 inhabitants result in similar changes in longevity biomarkers observed in animal studies? Yes, it did.

On average the Biosphere 2 inhabitants reduced their body fat, weight, basal-metabolic rate (the number of calories used at rest), and body temperature, and decreased their levels of insulin, blood sugar, cholesterol, triglycerides, and blood pressure. After reviewing and evaluating these health data, the researchers concluded that the Biosphere 2 calorie-restriction experience was not detrimental to health. In fact, it was health enhancing given the positive measurements in the biomarkers of health and longevity.

More Proof from Current Research on Humans

While this book was being written in 2006, an important research study was completed that contributed additional evidence to the calorie-restriction life span model. This research was funded by the National Institutes of Health and headed by a leading researcher in the field, Dr. Leonie Heilbronn. The team of researchers sought to determine what the effect of following a calorie-restricted diet for six months was for a group of forty-eight healthy but sedentary men and women.

Dr. Heilbronn and coworkers divided up the men and women into four groups, with each following a different dietary program. The control group ate a weight-maintenance diet. The other groups ate calorie-reduced diets. For example, one group ate a diet that was restricted by 25 percent. Another group ate a very low calorie diet for about eight weeks, followed by a 15 percent calorie-reduced diet until the end of the study. The remaining group followed a 15 percent calorie-reduced diet for the full term of the study. This last group also exercised and increased caloric intake slightly on exercise days.

After just six months of calorie-restricted diets, the results of this landmark study revealed a beneficial effect on some of the biomarkers for increased life span. These favorable results included reduced body weight and body fat, lowered insulin levels and body temperature, and less DNA fragmentation, which means less DNA damage. That means researchers found many of the same changes seen in long-term calorie-restriction studies in other mammals.

THE HUMAN GENOME PROJECT

While all that research was going on, we were putting this raw science into practice and determining ways to actually produce some of the health and longevity benefits without calorie restriction. That was made possible from our in-depth understanding of how genes work. You see, in addition to decades of research about how calorie restriction can increase life span in animals, another area of research important to reducing disease and increasing longevity involves mapping out the human genome. (The term "genome" is a catch-all that refers to all the genetic information found on all your chromosomes. See the "Common Genetics Terminology" inset on page 10, which explains a few key genetics terms that will be useful when reading this book. For more detailed definitions of many gene-related terms, see the Glossary at the end of the book.)

Humans have twenty-three pairs of chromosomes, one set inherited from each parent. As a result of the human genome project, we now know that about 30,000 genes are stored on these sets of chromosomes, which are present in each of our cells. With this information, we were able to gain important insights leading to unlocking the secrets of longevity and health. This research continues to reveal a connection between the expression of our genes and characteristics of health or the onset of disease. Also, research has confirmed that damage to our DNA and poor function of our DNA are the major reasons behind failing health and rapid aging.

What Is DNA?

We are all familiar with television shows like *CSI* and *Law & Order,* where an individual is identified by taking a sample of their DNA. This new technology lets us identify the most fundamental information that makes each one of us unique—the code stored in our DNA.

DNA, or deoxyribonucleic acid, is the chemical inside the nucleus of each cell that carries the genetic instructions for making that living organism. Aging occurs in DNA as newly forming cells look to their predecessors for their identity, in a process analogous to photocopying. (See Figure 1.2 on page 11.) Over time, factors such as environmental pollution, dietary behavior, lifestyle issues, and genetic inheritance erode the cellular infrastructure. As the quality of the new cells degrades, the molecules, nutrients, and chemicals circulating in and among the cells become deficient. This causes a breakdown in functions throughout the body and, over time, promotes aging.

COMMON GENETICS TERMINOLOGY

DNA (deoxyribonucleic acid): This double-stranded molecule holds encoded genetic information. It is held together by weak bonds between base pairs of nucleotides. The four nucleotides in DNA contain the bases adenine (A), guanine (G), cytosine (C), and thymine (T). In nature, base pairs form only between A and T and between G and C; thus the base sequence of each single strand can be deduced from that of its partner.

Chromosome: A chromosome is made from two strands of DNA. This self-replicating genetic structure of cells contains the cellular DNA that bears the genetic information in sets called genes. Humans have forty-six chromosomes, which are divided into two sets of twenty-three, one set inherited from each parent. Higher animals, including humans, have a number of chromosomes whose DNA is associated with different kinds of proteins.

Genetics: The study of inheritance patterns of specific traits.

Genetic polymorphism: Difference in DNA sequence among individuals, groups, or populations. For example, the genes for blue eyes versus brown eyes.

Genome: All the genetic material contained in all the chromosomes of a particular organism.

Genomics: The study of genes and their function.

Human Genome Project: Research and technology development aimed at mapping and sequencing the genome of human beings and certain model organisms.

What the newest scientific evidence has revealed from our research, together with discoveries from the Human Genome Project and other independent research, is that aging can be controlled. We now know enough about the biological process to make aging healthier—so that each of us can age more optimally on an individual level. You will learn more about this in Chapter 4.

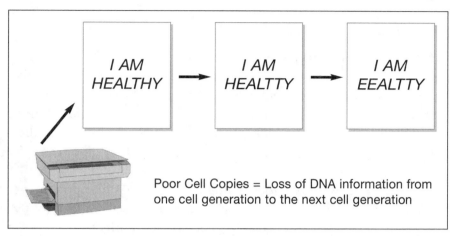

Figure 1.2. The effects of DNA damage over time.

Environmental Factors and Lifestyle Affect Genes

As calorie-restriction research revealed, the expression of genes is influenced by environmental factors and our lifestyle, which can affect our growth and development, start chronic diseases, speed the aging process, and cause premature death. That means gene duplication and the synthesis of proteins we need to live (which are the primary substances our cells make from information stored in our genes) and modifications of these substances are all affected by environmental factors and our lifestyle. For example, the primary substance our cells make from the information stored in genes is protein, which is responsible for many functions in our bodies. Therefore, constantly breathing high levels of air pollution will adversely affect the proteins responsible for many functions in our bodies.

The functioning of each of our cells depends on its ability to synthesize proteins in a timely and orderly fashion. These proteins have different but equally important functions in the body, so how well your cells keep making them is vital. But with aging, the adverse effects of the environment, poor nutrition, and bad lifestyle habits can cause malfunctioning in your cells' gene function. This leads to premature aging and the development of all sorts of age-related diseases. But with our new gene-smart information, you can control your health at the most basic cellular level and live healthier and longer. We call this know-how *Personal Genetic Health.*

PERSONAL GENETIC HEALTH

What sets our research apart is that it focuses on providing the longevity benefits of calorie restriction at the cellular level and on improving other genetic-based factors using nutritional programs and products. This led to our developing the new concept of health we call Personal Genetic Health (PGH). Our PGH Gene Makeover approach emphasizes what is best for each individual's total health and longevity based on what is best for their unique set of genes and cells.

For example, when we look at the effects of PGH Gene Makeover lifestyle or environmental change on gene activity, we can identify five key groups of health control genes that are markedly affected by how we live. Coincidentally, these are the same gene groups affected in calorie-restriction experiments in animals. These health-related genes control such important bodily functions as:

1. Glucose, or blood sugar regulation

2. Inflammation within the cells

3. Antioxidant levels

4. Methylation, or gene activity

5. DNA repair

Our Personal Genetic Health medical approach is pioneering. We know that the ability to identify the functioning of these health control genes and then influence their favorable expression can lead to living a longer, disease-free life. This new frontier of PGH Gene Makeover health-care includes gene testing followed by nutrition and lifestyle programs, an approach carefully designed to increase life span and decrease the risk of developing certain diseases.

How is this possible? Through testing known as gene SNP (single nucleotide polymorphisms) analysis. We will discuss this topic throughout the book, but here's a short overview.

What Is SNP?

We now know from the Human Genome Project that the DNA of any two

individuals is about 99.9 percent identical. Yet everyone, except for identical twins, is genetically unique. The distinguishing difference lies in the
0.1 percent. These small but meaningful differences are due to tiny differ-
entiations in some of our genes. Geneticists refer to these differences as
single nucleotide polymorphisms, or SNPs, which is pronounced "snips."

Our DNA can be thought of as a long chain, with each link represent-
ing a molecule. As it turns out, these long DNA strands are made up of
repeating sequences of just four molecules: adenine, cytosine, guanine,
and thymine. These four molecules are actually related in their structure,
so scientists often refer to them by their group name, nucleotides. All of
the instructions stored in our entire human genome are made up of difference sequences of these four molecules. When some genes that should
theoretically be identical are different due to one of these molecules, it is
referred to as single nucleotide polymorphism. (See Figure 1.3.)

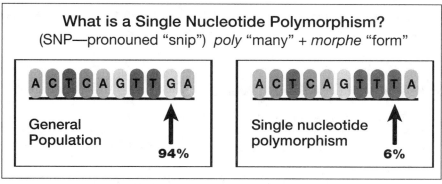

Figure 1.3. When a gene that is the same in most people is different in
one individual it is called a single nucleotide polymorphism (SNP).

In practical terms, our study of the behavior of DNA has revealed that
SNPs are associated with increased risk of certain diseases and the symptoms commonly associated with aging. Genetic research has also revealed
that DNA can be activated in a positive or negative fashion based upon
the quality of the environment in which it is placed. Therefore, the SNP
test, described in detail in Chapter 7, is an essential tool in determining
the most effective components of a PGH Gene Makeover Age Management Program personalized for a specific individual.

Experiencing the Benefits of Extending Life Span and Health Span without Calorie Restriction

The genes involved in controlling health and longevity, as documented by our gene SNP analysis, can be grouped into five main categories related to the bodily function groups listed above. Only recently did we learn how to measure our human gene potential in these five "health control" gene groups. This recent offshoot of the Human Genome Project has revealed the types and quantities of nutrients or nonprescription supplements that can be used to closely mimic or reproduce the effects of calorie restriction on the cellular level. We can accomplish this today without actually cutting our calories as severely as we saw in the original research projects.

Despite the earlier research some people still do not believe that calorie restriction is the answer to extending human life span and health span. The good news is that based on additional research and our experience, each of the health control gene categories related to disease and longevity can be beneficially controlled with nutritional intervention rather than calorie restriction. Food and supplements can:

- Improve insulin sensitivity and blood sugar control.

- Reduce and control inflammation.

- Improve antioxidant protection and function.

- Turn off disease-causing gene expression and turn on health-promoting genes.

- Promote DNA repair.

DNA contains a code that controls the five health-related processes in each one of the 100 trillion cells in our bodies. As noted earlier, each cell has about six feet of DNA. (See Figure 1.4.) When we calculate the quantity of DNA within each of our bodies—by multiplying six feet times 100 trillion—we find the staggering figure of billions of miles of DNA. The sheer quantity of DNA and the vital code it carries makes it incredibly important to all aspects of our being. And now our revolutionary Personal Genetic Health approach pinpoints which lifestyle and environmental factors are good for you and which you should avoid.

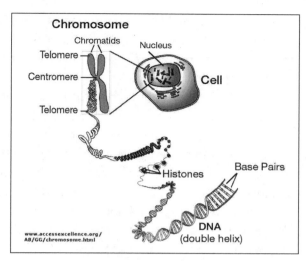

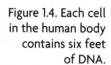

Figure 1.4. Each cell in the human body contains six feet of DNA.

How Is DNA Damaged over Time?

The Unified DNA Damage Theory of Aging, proposed by Dr. Giampapa, explains that various kinds of damage to DNA cause damage to our genes, as well as initiate changes to our gene expression, which accelerates aging. There are five key cellular processes related to aging and DNA damage. **The Positive Health Genes repair damaged DNA at the cellular level to maintain optimal cell function.** Genes control these cellular processes that regulate clinical aging:

Methylation: Turning specific genes on and off (or regulating gene expression); generally related to the heart and vascular health.

Inflammation: Response to cellular injury that includes swelling and pain; generally related to bone and joint health.

Glycation: Blood sugar effects on proteins and body fat; generally related to loss of youthful body shape and composition.

Oxidation: The production of free radicals; generally related to damage to our DNA and genes.

DNA Repair: Restoring DNA to its optimal state.

Each of these processes is controlled by a specific group of genes. If these cellular processes become damaged and as a result are poorly controlled over time, or if they were suboptimal because they were based on poor genetic inheritance to begin with, DNA damage occurs at a more rapid rate. As the processes become less efficient at the cellular level, we age more rapidly.

These five health-controlling processes maintain our health, but they also contribute significantly to the aging process at the cellular level. Dr. Giampapa's theory utilizes the latest genetic research, so we can now measure the activity or efficiency of these five processes in each person and use this information to promote a healthy, long life.

Since everyone is different even at the cellular level, we can now define what aging is for each individual, and what supplements are needed to slow the process for that specific person. After all, it is not what the average person needs that is important, but *what we need as individuals* that can make all the difference in how we age and the quality of our health.

The Anti-Aging Breakthrough

The revolutionary PGH Gene Makeover breakthrough is based on the five principle health-controlling cellular processes discussed above. It focuses on the new science of gene SNPs and testing the genes involved in these processes. Once your personalized information is determined we can recommend specific vitamins, minerals, or nutraceuticals that should be used *by you, for you* to keep your genes functioning at their most optimal levels. Once you start the program you can monitor your DNA to see if you are aging more optimally by using a small home-based urine test. Any worry about how you are aging will then be gone.

This book will show you the importance of providing the best and most optimal environment for your genes. But your genes alone do not determine how you age or your level of health. That's why this book will also discuss why your lifestyle as well as your beliefs and emotions are essential parts of your new road to successful health and longer life.

It is our ultimate goal to give each of you the ability to personally direct and monitor your own health—to be more proactive and responsible for the quality of your life and less dependent on the health care system, which focuses on the symptoms of disease and of aging rather than their causes.

THE IMPORTANCE OF OPTIMIZING THE AGING PROCESS!

Here are a few facts that emphasize why we need to focus on adopting the Personal Genetic Health anti-aging lifestyle. (See Figure 1.6.)

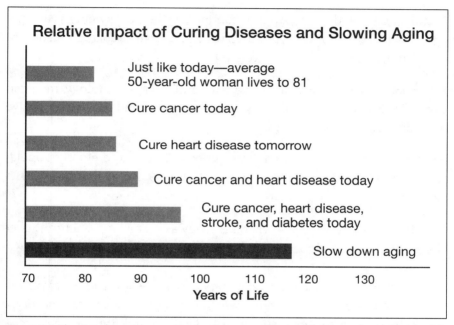

Figure 1.6. Slowing the aging process may be more effective for extending a healthy life span than the combined effect of curing cancer, heart disease, strokes, and diabetes.

Fact: Every second a baby boomer turns fifty in the United States.

Fact: In 1995, the baby boomer generation saved $15 trillion for retirement. However, if the same disease patterns of aging continue, they will actually need $184 trillion to maintain their health as they age. There is not that much money on the planet.

Fact: Current medical anti-aging technology has the ability to provide at least 50 percent of the 76 million baby boomers with a healthy life expectancy of 90 to 100 years.

Fact: If the health span—that is, maintaining a healthy, productive condition without disease—of every American could be extended just one year, the United States economy would save $1-3 trillion.

Fact: About 14 percent of the federal United States budget is spent on geriatric medicine, and this expenditure continues to grow.

Fact: If the same disease patterns continue in the next twenty years, the health care system of the United States will become bankrupt! (Laurence Kotlifoff, Economist, Brown University)

Fact: Once we reach age sixty-five with anti-aging technology, men can expect to live an additional 15.8 years, whereas women can expect to live an additional 17.6 years.

Fact: According to the United States Census Bureau, the elderly population in the United States will grow to between 59 and 78 million by 2030, one fifth of the total population.

Fact: By the year 2050, there will be as many as 2.2 million Americans over the age of 100. The global population will total 9.3 billion.

You can begin to change the quality of your health and how you age by reading this book. It will empower you to change your life. This book will help you master your own health destiny—and not be a victim of your genes. It will give you the information to change your beliefs, which will also help you alter your health destiny. It will show you that the secret to your future health is *to maintain your DNA in optimal health* by measuring it and limiting damage to it. This book will show you the most recent scientific evidence that the next best thing to the fountain of youth and the magic pill you've been searching for is already here!

2

The First Secret Revealed: If You Know How to Keep One Cell Healthy, You Can Do It for 100 Trillion of Them

Errors, like straws, upon the surface flow;
He who would search for pearls, must dive below.

—JOHN DRYDEN IN *ALL FOR LOVE*

Take a close look at yourself in the mirror, and compare the way you look with photos from your younger days. If you are like most people, some big differences will be apparent. This is probably one of the reasons that motivated you to purchase this book—to find out about advancements in medicine so you can do something about aging and age-related diseases.

You are probably wondering why these signs of aging are occurring and what you can do about them. To answer this important question, let's take a closer look at what is actually going on in your body to understand what you see in the mirror. When we understand the controlling factors, we can then take actions to control our aging destiny. For the most important answers, we must go deep within our bodies.

What the most recent advancements in science show is that we must use our most powerful zoom lenses to focus on what's happening inside our *cells* to understand how to control the aging process. Only then is it possible to determine what we can do about aging and age-related diseases. The good news is that this is what we have done for the past decade. We have new answers to help you control your health in a way never before possible.

This quest for medical understanding of what can be done to control aging and age-related diseases is what led us to develop the Personal Genetic Health Gene Makeover. We're all aging—our patients, our families, friends, colleagues, and you. And like you we share the desire to age gracefully and live longer, healthier, more satisfying lives.

That's what drove us to find out why some people look younger than others, why they live longer, or why they are free of the common illnesses of aging. After years of searching, we found the answers we were looking for hidden deep inside each cell in our bodies. The best part of what we discovered is that with this new information, we can take actions to increase our life span and ward off the common diseases of aging.

THE ANTI-AGING ANSWER REVEALED

The secret of aging well and staying healthy is quite simple: *When you know how to keep one cell healthy, it's the same process to keep 100 trillion cells healthy.*

1. Decrease DNA damage and increase DNA repair

2. Decrease free-radical damage

3. Overcome inherited genetic weaknesses

If we look at aging on the cellular level, one cell at a time, it becomes possible to accomplish what in the past was considered impossible: We can alter our genetic fate. To begin, we must consider two points. First, we need to keep in mind that the information needed to accomplish this goal has really only been available for a few years, since the Human Genome Project was completed. Second, so much information was revealed by the Project that we didn't know what to do with all of it until it was sifted through, reviewed, and analyzed for practical applications. Only then could we take these new medical insights and develop the Personal Genetic Health Gene Makeover Program, which you can benefit from today. What's more, PGH Gene Makeover will change all aspects of how medicine is practiced in the future.

The Basic Genetic Blueprint

What we now know from the Human Genome Project is it takes about 30,000 genes to create the blueprint that makes a human being. Each gene contains instructions about how to make different substances we call proteins. These proteins are then used to make molecules. These molecules cannot function alone, so they are assembled in different ways to make different cells in your organs and other tissues, like muscle and bone. Each human being is a collection of 100 trillion cells working together in harmony. When your cells are not working correctly and in harmony, the result is premature aging and the plethora of health problems we associate with aging.

It is interesting to note, however, that at any one time only about 10 percent of your genes are turned on or actively functioning to make proteins. This is an important concept to keep in mind. That means about 3,000 genes out of 30,000 are working at any specific time in our life.

Why is that important? This is a loaded question, but one part of the answer is that our genes get activated at different times during the aging process and under different circumstances. During our youthful growth years, specific genes work to keep us youthful. Then, as the years go by, the function of these "youth" genes declines, and the so-called "old-age and disease" genes get turned on. (See Figures 2.1 to 2.3.) But it doesn't have to be that way. Our new understanding of how our cells work has enabled us to develop ways to keep people more youthful and healthier longer from the genetic level of our cells outward.

Positive Aging • Childhood Profile

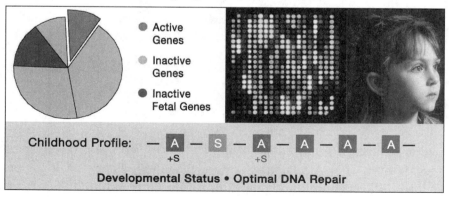

Figure 2.1. Changing gene expression during aging: youth.

Maintenance Aging (Age 21–30 +) • Adult Profile

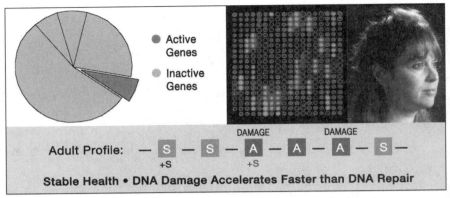

Figure 2.2. Changing gene expression during aging: after age 30.

Negative Aging • Accelerated Disease and Rapid Aging

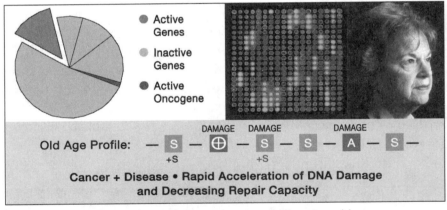

Figure 2.3. Changing gene expression during aging: old age.

Positive versus Negative Aging

Positive aging is what we need to strive for. The good news is we can influence the aging process by the life we choose to lead—by selecting positive-aging foods, taking supplements based on new gene tests, doing specific kinds of physical activity, maintaining desired body composition, and even thinking healthy thoughts each day. To accomplish these anti-

The First Secret Revealed

aging goals, we must better understand the basic processes of aging, especially the ones that are in our control.

First, it is important to realize that there is a period in our lives when we want to age or develop, and specific groups of genes are activated to accomplish this. This period is from infancy to adulthood. Positive aging allows us to develop and mature into healthy young adults, with all our physical and mental abilities intact and at their peak.

Second, within the youthful positive-aging genes are a number of genes involved in keeping each cell healthy, in addition to helping us develop into young adults. Once we reach young adulthood, ages twenty to thirty, we have attained a state in life where we would ideally like to stay as long as possible. This is the state of ideal health and optimal function of each of the trillions of cells in the body. We never feel better, think faster, or heal quicker than at this time of life. Our youthful and health maintenance genes are working 24/7 to keep us young and healthy. The trick is to keep our youthful genes turned on as long as possible and keep our aging-related genes turned off.

The Life Span Secret

During young adulthood, the active 3,000 genes are involved in healthy maintenance, optimizing us for reproduction and caring for children. We can stay out late, party through the night, even abuse our bodies to a certain point with bad lifestyle habits, and still stay well, keep thin, and maintain our energy levels. Researchers estimate 30,000 to 70,000 metabolic events damage our DNA every day. When we are young, our cells can repair the majority of this daily damage.

The primary reason this repair can happen at any point in our lives depends on key genes that repair the damage that we inflict on our DNA. The DNA repair genes are the most important genes we have. Keeping our DNA repair genes working well as we age is essential for optimal aging. (See Figure 2.4.) Turning our attention to the animal kingdom reveals an interesting fact about the importance of DNA repair for longevity. Animal species with greater DNA repair ability have longer life spans. If we look at human diseases that affect certain people's ability to repair damage to their DNA, like Werner Syndrome, Hutchinson-Guilford Syndrome, or

Xerodema Pigmentosa, we know that these individuals have much shorter life spans and health spans. So, while several factors control the aging and degenerative disease process, maintaining the ability to repair your DNA is chief among them. Many foods and supplements play an essential role in keeping DNA healthy.

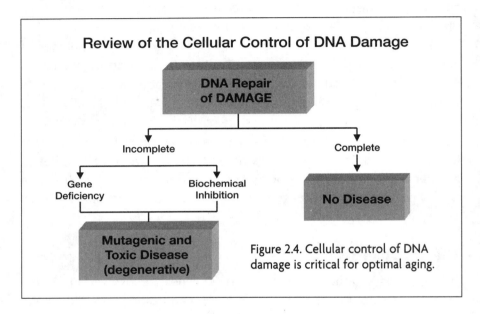

Figure 2.4. Cellular control of DNA damage is critical for optimal aging.

Negative Aging: The Problem Is Inefficient DNA Repair

Here is another vital statistic. At around age thirty, we begin to see and feel a difference in how we look and the quality of health we experience. We feel the beginning of the downside of the aging process, because our DNA repair genes can no longer repair the damage to our genes as fast as it occurs. This is the point where *negative aging,* or what we normally think of as aging, actually begins.

While this is a natural process, we can stack the odds of aging optimally—and as slowly as possible—in our favor, with healthy lifestyle factors that maintain maximum DNA repair and reduce DNA damage. For example, as you will read in Chapter 7, all other factors being equal, the supplements you take can be a significant part of your anti-aging lifestyle. When you adopt a positive anti-aging lifestyle—as opposed to a negative lifestyle of inactivity, all you can eat, anything goes—you can live a longer and healthier life. A quick look at some extreme instances of premature aging and causes of degenerative disease will help illustrate this point.

Lessons Learned from Progeria, the Rapid Aging Disease

Every now and then, in movies or on television shows, you'll encounter the story of someone who ages fast. There's even a comedy where the main character ages so fast he reaches the size of an adult in grade school. Though it seems like science fiction, this condition is real, and we have learned important lessons by studying this very rare genetic disorder called progeria (or Hutchinson-Gilford progeria after the scientists who discovered it). One of the key factors causing this genetic-based, rapid-aging disease is a poor DNA repair mechanism. This is caused when the telomeres that cap the ends of our chromosomes are shortened, which limits the number of times the cells can divide before they die.

Progeria, whose name comes from *geras,* the Greek word for old age, is estimated to affect one in 8 million newborns worldwide. As newborns, children with progeria usually appear normal. However, within a year, their growth rate slows, and they are soon much shorter and weigh much less than others their age. So the positive-aging process stops early and the negative-aging process begins prematurely.

Children with progeria develop a distinctive appearance characterized by baldness, aged-looking skin, a pinched nose, and a small face and jaw relative to head size. They also often suffer from symptoms typically seen in much older people: stiffness of joints, hip dislocations, and severe, progressive cardiovascular disease. Some children with progeria are given coronary artery bypass surgery and/or angioplasty in attempts to ease the life-threatening cardiovascular complications caused by their progressive atherosclerosis. However, some other features associated with the normal aging process, such as cataracts and osteoarthritis, are not seen in children with progeria. Tragically, there currently is no treatment or cure for the underlying condition these children have. Death occurs on average at age thirteen, usually from heart attack or stroke.

What causes progeria? This question perplexed the world of medicine until just recently. In 2003 scientists at the National Human Genome Research Institute, working together with colleagues at the Progeria Research Foundation, the New York State Institute for Basic Research in Developmental Disabilities, and the University of Michigan found the answer. This condition is caused by a mutation in a single gene known as *lamin A.*

While progeria is a genetic-based disease, the parents and siblings of children with progeria are, paradoxically, virtually never affected by the disease. So it appears that something else is in play beyond passing the disease through the generations. After some checking, the researchers determined that the genetic mutation appears to occur in the sperm in nearly all instances.

It is remarkable that nearly all cases of progeria are found to arise from the substitution of just one base pair among the approximately 25,000 DNA base pairs that make up the lamin A gene. If we take a closer look, the lamin A gene makes two proteins, called lamin A and lamin C. These proteins play a vital role in stabilizing the inner membrane of the cell's nucleus. The genetic mutation that causes Hutchinson-Gilford progeria signals the lamin A gene to produce a defective form of the lamin A protein, which makes the nucleus and its DNA contents more susceptible to damage.

This abnormal lamin A protein appears to cause destabilization of the cell's nuclear membrane in a way that is particularly harmful to tissues routinely subjected to intense physical force, such as the cardiovascular and musculoskeletal systems. The researchers also detected that different mutations in this lamin A gene are responsible for at least a half-dozen other genetic disorders, including two rare forms of muscular dystrophy. In addition to its implications for diagnosis and possible treatment of progeria, the discovery of the underlying genetics of this model of premature aging has helped to shed new light on the aging process.

Telomeres: Another Measure of Health

Progeria patients are born with shortened telomeres, the rod-shaped structures found at the end of a chromosome. These specialized structures are involved in the replication and stability of DNA molecules during cell division. (See Figure 2.5.) It is thought that each time a cell replicates, the telomeres are shortened. After many cell divisions, the telomeres

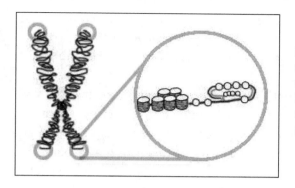

Figure 2.5. Telomeres are specialized structures on the ends of chromosomes that are involved in the replication and stability of DNA molecules during cell division.

become shorter and shorter, eventually triggering the signal for the cell to die. Because progeria patients have cells with short telomeres, this shortens the life span of their cells and contributes to their rapid-aging syndrome.

In the 2005 *Journal of the American Geriatrics Society,* scientists reported that maintaining normally-functioning telomeres is important for prolonging the healthy function and life span of our cells. And Dr. E. Woodring and co-authors of a 2005 scientific report entitled "Telomere Biology in Aging and Cancer" pointed out that other natural forces can cause telomere damage—namely, oxidative stress.

Premature Aging Factors That Are Under Our Control

While the progeria story illustrates the powerful premature-aging effects that *just one gene mutation can have in humans,* imagine living a lifestyle that causes damage to hundreds of your youth genes and prematurely turns on your aging and disease genes. High levels of chronic stress can also cause telomere shortening, as documented in a 1989 landmark study by Dr. Elizabeth Blackburn. Are you starting to see the big picture on the gene, DNA, and aging relationship?

Smoking, eating the wrong foods or not eating the right foods, lack of exercise, the polluted air we breathe, the unclean water we drink—are all behaviors under your control that affect the way your genes work as well as your health and longevity. The biggest premature killers, like cancers, cardiovascular diseases, type 2 diabetes, and obesity, are mainly the result of living a lifestyle that turns off your youth genes, causes more DNA damage than can be repaired, and turns on your disease-causing, age-accelerating genes. By living the PGH Gene Makeover lifestyle, you can reverse this negative progression and increase your life span and health span.

You will recall that up until age thirty, your positive health genes are activated and work optimally. After age thirty a different story begins, but you can have a hand in altering it for the better. From this age on, our cells can't repair their DNA as fast as the damage occurs. This means our genes can no longer function efficiently to maintain our health and the quality of life we would like to enjoy. Our cells can't produce the proteins they need to maintain healthy structure and function.

POSITIVE AND NEGATIVE HEALTH GENES AND THE CONSEQUENCES OF NEGATIVE AGING

After age thirty most of us start to experience the consequences of poor gene functioning and our vibrant health slowly declines. We start to gain weight as our metabolism slows, because our cells are not producing as much energy as they did when we were younger. We notice aches and pains in our joints because our tissue repair can't keep up with tissue inflammation and damage in these parts of the body. Our skin begins to sag and wrinkle because the damage to the collogen and elastin in the skin caused by internal and external substances is not being repaired fast enough anymore.

Other hidden signs of aging also occur inside our bodies, including damage to our arteries, inflammation of our cells, and a decline in both the number and function of our brain cells. Eventually, the positive health genes decline so much that we look, feel, and act older than we should.

The Positive-Health Genes

Studies show that certain genes, or groups of genes, allow mammals to live up to 40 percent longer. The genes involved in keeping us healthy and functioning optimally include the same group of genes that responded favorably in the calorie restriction experiments discussed in Chapter 1. We now know that we can do things to keep these important genes working at their positive, active level much longer, starting with something as easy as taking the right supplements and neutraceuticals.

Using our PGH Gene Makeover approach, we have grouped these positive genes into five main categories presented in Table 2.1.

Table 2.1. Summary of the PGH Gene Makeover Positive Gene Groups

1. Methylation genes help activate and deactivate, or switch on and off, other genes in the correct order. Methylation is the addition of a carbon-hydrogen molecule to a methyl molecule, which controls the masking of specific regions of DNA and the unmasking of others. This alters how our genetic switch turns from on to off or from off to on and determines which genes are active or inactive. Cells are healthiest when methylation occurs properly. The process is similar to placing your finger over specific holes in a flute to create different notes to produce a harmonious melody. The evidence emerging from recent research strongly suggests that brain aging and possibly cancer and cardiac aging are, in part, consequences of altered methylation patterns.

2. Inflammation genes are involved in inflammation in the cells, bones, and joints. Certain substances in the body trigger inflammation gene activity causing cell and tissue damage, swelling, pain, and dysfunction.

3. Glycation genes are involved in blood sugar control and body fat levels. Elevated glucose levels, insulin surges, and poor insulin receptor sensitivity cause the cross-linking of proteins at the cellular and genetic levels. This directly affects gene activity or expression and protein synthesis. Glycosylation (sugar chemical reaction) of immune system molecules called immunoglobulins modifies their function and may contribute to autoimmune reactions and poor immune function in general.

4. Oxidation genes regulate antioxidants needed to protect DNA from damage by free radicals. Oxidative stress is the amount of free-radical damage produced outside and inside the cells. This directly affects genetic structure and function as well as cell membrane and internal cell function.

5. DNA repair genes are the most important genes because they regulate repair of DNA damage. These genes are key to keeping every cell in the body healthy and working at their youthful prime. Damage to DNA is like having a glitch in a computer software program that makes it difficult to interpret information and execute a task. This important topic will be covered in Chapter 3.

The Causes of DNA Damage: How Aging Begins

You are probably thinking that with 30,000 genes in each cell and trillions of cells actively dividing each day, something must go wrong with the DNA duplication process to cause the changes associated with aging. And it does. Each day structural alterations occur in our DNA. Remember that each day 30,000 to 70,000 events damage DNA in each cell. These structural alterations are referred to as mutations. Most of these mutations have no effect and do not cause us any harm. In fact, many of the mutations that occurred through the ages of human development were actually beneficial. However, sometimes mutations can cause problems.

Here are some of the major mutation-related problems that can happen to the DNA in your cells. Any one of these five mutations can cause disease or premature aging if left unrepaired:

- *Deletion* occurs when a chromosome loses a piece of DNA. Deletion of a gene or part of a gene can lead to a disease or abnormality. (Figure 2.7.)

- *Duplication* occurs when one or more of any piece of DNA, including a gene or even an entire chromosome, is copied onto the same strand (where it is already present). (Figure 2.7.)

- *Inversion* is when a portion of the chromosome breaks off, is turned upside down, and is reattached in the incorrect position. (Figure 2.7.)

- *Insertion* occurs when a portion of a chromosome that has broken off forms a circle or ring inside the cell. (Figure 2.8.)

- *Translocation* occurs when a large segment of DNA from one chromosome breaks off and attaches to a different chromosome. (Figure 2.9.)

What Causes Cells to Age—Individually or at the Mass Level?

The answer to that question is simple: damage to the DNA that controls a particular group of positive-health genes. But where does the damage come from? It comes mainly from free-radical compounds released by the food we eat as we digest the carbohydrates, proteins, and fats it contains. The unhealthy twenty-first-century environment is also a major contributor to DNA damage. In fact, humans have never been exposed to so many toxins as we are today, given all the artificial chemicals in the soil that end

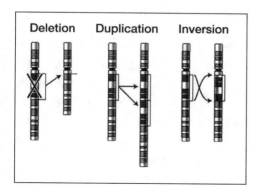

Figure 2.7. Deletion, duplication, and inversion are three kinds of DNA mutations that can lead to disease.

Figure 2.8. A fourth type of DNA mutation called insertion occurs when part of a chromosome breaks off and forms a ring inside the cell.

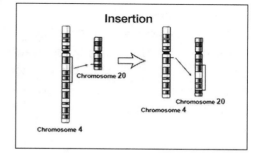

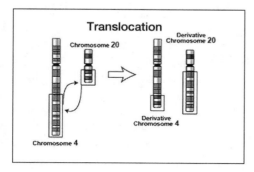

Figure 2.9. A fifth type of DNA mutation called translocation occurs when part of the DNA from one chromosome breaks off and migrates to another chromosome.

up in the food we eat and all the exhaust and industrial smoke fumes in the air we breathe. These environmental insults include cell phone radiation, which disrupts the way DNA works and may lead to bad DNA duplication. (The health consequences of environmental toxins are covered in Chapter 4.) The environment we now live in is far from the optimally healthy one we genetically evolved from over the past million years. We live on a very different planet than we were originally intended to live on.

Activation of Negative Health Genes

At the critical point in our lives after thirty, the accumulation of free-radical damage and environmental toxins can activate negative genes—the genes that cause cancer, Alzheimer's disease, arthritis, type 2 diabetes, and cardiovascular diseases. For instance, cancer-causing genes are called *oncogenes,* and they have the ability to begin the transformation of normal cells into cancer cells. It is interesting to note that there are also genes that can suppress tumor growth called *tumor suppression genes.* But under conditions of negative aging, a tumor suppressor gene can mutate and fail to keep a cancer from growing.

For example, genes identified as BRCA1 and p53 were among the first tumor suppressor breast cancer genes to be identified. The p53 gene normally regulates the cell cycle and protects the cell from damage. Mutations in this gene cause cells to develop cancerous abnormalities, leading to the development of breast cancer. In fact, mutated forms of these genes are believed to be responsible for about half of all cases of inherited breast cancer, especially those in younger women.

THE FIRST ANTI-AGING SECRET REVEALED

If you know how to keep one cell healthy, then you can keep many cells healthy and age more optimally. So the first anti-aging secret involves the ability to perform the following positive aging behavior at the cellular level:

1. Limit or decrease free-radical levels. Free radicals are the most likely cause of DNA damage. Scientists have shown that we can do this now by taking certain supplements, eating the right foods, and exercising in the right way.

2. Increase DNA repair. Loss of the DNA repair mechanism within cells is a major cause of premature aging and age-related diseases. Scientists have shown that we can increase DNA repair with positive-aging lifestyle practices.

3. Use positive-health genes to their optimal capacity. Our positive-health genes can be tested for efficiency, and then we can adjust our

lifestyle and take supplements to sustain those genes. Scientists have shown that we can do this now.

START A PGH GENE MAKEOVER PROGRAM TODAY TO PREVENT NEGATIVE AGING

In the following chapters, you will learn more about the Personal Genetic Health Gene Makeover, which combines the latest findings about genetics, nutrition, vitamin and mineral science, and herbology as well as anti-aging medicine. The PGH Gene Makeover Program includes a convenient way to test your positive-aging genes so you will know exactly what is going on inside your body as far as DNA damage and free-radical levels. Then, using the results of your test, you can determine the supplements you should take to maintain positive health and reduce your risk of developing degenerative diseases.

Here's a real-life example of how the PGH Gene Makeover Program works. Mrs. R., a seventy-two-year-old female, approached me after a public lecture. She said, "I loved your lecture, and it seemed to make so much sense, but what can I do? Geez, I'm seventy-two, and I must have already suffered so much DNA damage. I have arthritis and a poor heart, and I cannot eat many of the foods you suggested because they make me feel bad and upset my stomach. I'm hopeless."

I suggested she try the PGH Gene Makeover Program under the guidance of her doctor. Six months later, I received an e-mail from Mrs. R.'s doctor, stating that she had experienced remarkable lessening of her joint pains as well as improvement in her ability to walk and exercise. In addition, she showed definite improvement in her skin quality as well as in her overall energy level.

For the first time, said the physician, he believed in the power of supplements. But he was convinced it was because the PGH Gene Makeover program was based on testing Mrs. R.'s genes first to determine her specific requirements. "I believe she responded well," he wrote, "because her body is getting what she needs as an individual and not generic vitamin and mineral supplements based on minimum daily requirements for the masses. I believe this is the most effective program I have ever seen for a patient because it really focuses on preventive medicine."

As this case study shows, the PGH Gene Makeover program is an approach that puts the practice of effective preventative medicine in your hands. This is indeed a major advance in modern medicine, one that will help you to increase your chances of living a longer, healthier life, so your golden years will be truly golden. You are the first generation to receive this information. Become proactive, and use it to change your life.

3

The Second Secret Revealed: The Last Thing You Want Is a Bad Photocopy of the Most Important Information in Your Life!

When you teach your son,
you teach your son's son.

—THE TALMUD

I n Chapter 2 we discussed the aging process at the individual cellular level, and why it's so important to keep our DNA and the genetic information contained within it as healthy as possible. This is especially true for the genes that control the five key processes regulating aging at the cellular level. We stressed that if we can control aging at the individual cellular level, then every one of the 100 trillion cells that make up our body will age more optimally.

But how does this work? It comes down to *making better copies of your cells* as they naturally wear out and are replaced. The last thing you want is a bad photocopy of the most important information in your life—your DNA.

In this chapter we'll review what happens normally as we age in a positive fashion, with the focus on the processes of cell replication or cell replacement.

CELL REPLICATION

As our cells wear out or are damaged on a daily basis, they make copies of themselves to constantly renew the body. When we are age thirty or younger, our cells tend to make perfect copies because our DNA has the ability to repair virtually all the damage it routinely suffers. These newly copied cells replace the old worn out and damaged cells and take over whatever function they were assigned to accomplish. Whether they are liver cells, heart cells, skin cells, or muscle cells, this cell renewal process is vital to keep the body functioning optimally. Most of these new cells come from the adult stem cells in our bone marrow. We'll discuss this later in this chapter.

However, after age thirty the damage the cells suffer is not repaired optimally, and this repair capacity starts to decline on a yearly basis as we get older. As a result of this decline in the quality of cell replication, each successive generation of cells is copied a little less perfectly than the previous one, and the genetic messages stored on our DNA gradually become less efficient and more defective in their function.

The aging-control genes we discussed in Chapter 2—those that control cell aging for each individual—also begin to work less optimally after age thirty. The meltdown of these vital aging-control genes speeds up the onset of diseases of aging, such as diabetes, bone and joint pain, increased body fat levels, and loss of muscle leading to obesity, as well as the signs we hate to see on our faces—lines and sagging skin. When this process is left unchecked, it leads to premature aging and development of the degenerative diseases above, which have become all too common in the twenty-first century. Why does this happen? The primary reason is that most people are living a lifestyle that causes problems at the cellular level. However, they can easily turn things around with a few simple lifestyle changes and by taking the right nutraceutical supplements, as discussed in Chapter 6.

The Photocopy Analogy: Good and Bad Cell Copies of the Most Important Information in Your Life

The progressive decline in the DNA copy process is associated directly with cell replacement. As noted in Chapter 1, it's analogous to what happens when you make a photocopy from another photocopy. Most people know what happens when you make copies of copies. The content of the copy being produced becomes less legible as it becomes more remotely related to the original. Applying this same concept to the copies of DNA being made in your body through the years, it's easy to see that the genetic information that was present on the original DNA is eventually completely lost. (Figure 1.2 on page 11 shows that DNA damage creates poor cell copies.)

Telomeres: The Cellular Clock

In Chapter 2 we discussed telomeres in relation to the premature-aging disease progeria. Now you need to know more about how telomeres—the small areas at the ends of the chromosome—function when cell copies are made. When each cell copies itself (according to an internal cell division clock) the cell is unable to copy the full length of the original telomere. The copied cell has telomeres that are slightly shorter than its predecessor. After approximately fifty to eighty cell divisions, the cell sends a signal to itself to stop dividing, and it eventually dies without replacing itself.

The shorter a telomere gets the older the cell is, and the less perfect its genetic information. One of the most important recent findings is that free radicals and other forms of oxidative stress work to shorten telomeres, causing cells to age much faster than they would otherwise. When this happens, the defective cells can either become cancerous or be destroyed by a mechanism that's also genetically controlled within the cell. The body cleansing mechanism that rids our bodies of poorly functioning or damaged cells is called *apoptosis,* or "programmed cell death."

Apoptosis: Cleaning House

It's important for apoptosis to function correctly to keep the body free from malfunctioning cells. However, all too often, this important dead-cell-cleansing process declines as we age. When the cell-cleansing process declines our bodies have a higher number of poor, unhealthy cells, making us more susceptible to developing tumors and cancers. This is the root cause of a plethora of age-related diseases and poor health.

There is a potentially unhealthy force associated with apoptosis that we need to know and do something about if we are to have optimal aging. This involves an inflammatory agent called *tumor necrosis factor* (TNFα).

This inflammatory substance triggers the control signal that activates apoptosis, which is usually an orderly process that occurs in your body billions of times each day. That's right—billions. But when your cells and TNFα are exposed to many free radicals from oxidative stress, their ability to function properly and turn on apoptosis automatically becomes impeded, reducing its vital function. This means TNFα can no longer promote apoptosis normally when exposed to free radicals.

When TNFα is active, it may also trigger another substance in cells called *nuclear factor kappa-B* (NF-κB), which can reduce apoptosis. The really bad news is that, when activated, NF-κB generates a series of events that promote aging, including increasing DNA damage, inflammation, and poor immune function.

This is an oversimplification of complex events, but it demonstrates how external and internal factors like chemicals and free radicals can throw off our delicate genetic balance and cause cellular chaos that leads to premature aging and disease formation. The good news is that research and our increasing knowledge in this area means we're able to do something about it by taking certain nutraceutical supplements you'll learn about in Chapter 6. For instance, one of the natural ingredients in these neutraceuticals is Cat's Claw, considered a miracle herb from South America. A component in Cat's Claw called AC-11 has the proven ability to reduce damage to DNA, help promote normal apoptosis, and stimulate healthy immune system function. AC-11 has been researched for over a decade at the University of Lund in Sweden, where these benefits have been documented.

THE ADULT STEM CELL FACTOR

What fuels this constant supply of new cells that replace the old, worn-out cells? All the latest research in cell aging and genetics is now focused on the important role of adult stem cells in cell renewal. We're not referring to the embryonic cells used to clone new cells, which are surrounded by moral and ethical issues. We're referring to the normal stem cells we all have in our organs and particularly in our bone marrow.

When we're young, we have a relatively large number of adult stem cells in many different body tissues, including muscle, bone, nerves, digestive system, and even fat cells. They are also sometimes referred to as *undifferentiated somatic cells* because they can be activated to turn into the same kind of cells as that of the tissues that surround them. A person would not live very long without a healthy reserve of these adult stem cells to replace the cells that die from damage or after reaching the end of their ability to reproduce and function properly.

As the body's cells get damaged or worn out, they are normally replaced by cells from our adult stem cell pool. Like the cells that came before, these new, healthy body cells continue the copying process again and again, making new copies to ensure whatever tissues they form are functioning at their best level.

However, as we age, both the number and the function of our body's adult stem cells start to decline due primarily to the DNA damage they suffer. Here's what happens to various systems as adult stem cells decline:

- Integument system: Skin, fat, collagen, elastin, and fat levels decline.

- Musculoskeletal system: Muscle mass decreases.

- Endocrine system: Hormones decline.

- Nervous system: Central nervous system (memory) and peripheral nervous system (autonomic) decline.

- Genitourinary system: Kidney function decreases.

- Gastrointestinal system: Malabsorption develops, leading to leaky gut syndrome.

- Immune system: Immune function declines, leading to bacterial and viral infections and autoimmune reaction.

So there are fewer and fewer adult stem cells to replace our body's cells as we age. Plus, aging cells begin to make poor copies. One anti-aging goal and treatment to strive for in the near future will be to make copies of our adult stem cells and give them back to ourselves. But how well they work will depend on how much DNA damage they have already suffered.

One of the key approaches to delaying the decline in the number and function of adult stem cells is to decrease the DNA damage they suffer.

That is possible today by following the PGH Gene Makeover supplement, nutrition, and exercise plans discussed in subsequent chapters in this book. While researchers are trying to figure out how to culture our adult stem cells so they can be inserted into our tissues, our best defense is to live a life that promotes healthy cell function all the time.

DNA IS THE ULTIMATE TARGET FOR ANTI-AGING THERAPY

By now you understand why it's so important to keep the DNA in each one of the 100 trillion cells of your body in excellent condition. The first key factor to remember about DNA is that the better your DNA health is the better your cell function is, and the better your DNA will be able to use the five key groups of positive-health genes that keep our cells functioning optimally, as we discussed in Chapter 2. The second key factor is that keeping your DNA healthy results in better cell copies and aids the DNA repair process through the cell replacement process. (See Figure 3.1.)

Now that your attention is focused at the cellular level, you probably have questions, such as: "How do I know how much DNA damage I have right now?" and "How will I be able to see if the level of damage is decreasing so I know I'm aging more efficiently?"

Successful Repair	Unsuccessful Repair
Radiation, toxins, poor diet, and environment damage DNA strand.	Radiation, toxins, poor diet, and environment damage DNA strand.
Repair enzymes and protein process in conjunction with CAEs repair DNA. No damage evident.	Repair enzymes and protein process fails. Damage evident.
DNA copies itself making accurate cell copies.	DNA copies itself making poor cell copies.
Healthy Cell and No Disease = Optimal Aging in Somatic Cells and Adult Stem Cells	Unhealthy Cell, Chronic Disease = Accelerated Aging in Somatic Cells and Adult Stem Cells

Figure 3.1. DNA repair is linked to accurate cell copies and optimal gene function.

In the past, anti-aging programs required a large number of very expensive blood tests to determine what supplements or medications and lifestyle changes would improve the quality of your life and optimize your cellular aging. Today, however, because we know that DNA damage rates and free-radical levels are so important, we can limit our testing to the following two key compounds and get an excellent idea of how much DNA damage has occurred or how efficiently our anti-aging program is working:

- DNA damage rates are measured by determining the level of a compound called 8OHDG.

- Free-radical levels are measured by determining the level of a compound called 8-EPI-PGF-2-Alpha.

Two Key Tests Determine Your Genetic Protective Factor

Several decades ago, when scientists first determined that certain rays from sunlight can damage the skin and cause skin cancer, a new term entered our vocabulary: *sun protection factor* (SPF). Today, everybody knows the meaning of SPF and the importance of reducing our exposure, especially for children, to the damaging ultraviolet (UV) rays from sunlight. That is a good example of how a complicated, esoteric scientific topic eventually became simple to understand. The same will be true someday for all the genetic terminology used in this book.

The scientific community has developed two simple tests everyone can and should take to determine the level of DNA damage caused by their lifestyle and environmental factors. With just a small sample of urine, we can now measure our DNA damage rates and free-radical levels. This is possible because damaged DNA pieces removed from the cells and excreted in the urine can be measured at the 8OHDG level. The free-radical levels are determined by the amount of damage suffered by the cell's sensitive outer layer, the cell membrane. This is measured by looking at levels of lipid damage or peroxidation (8-EPI-PGF-2 Alpha).

We call these two values your *genetic protection factor* (GPF). (See the "Genetic Protection Factor" inset for more details.) When you get your test results, you should know that the lower your GPF values are, the better you are aging at the cellular and genetic level, and the better the quality is of your cell copies. The details of this and other tests are discussed in Chapter 7. What you may find most interesting about periodic PGH Gene Makeover testing is that it enables you to measure the positive-aging effects of your supplement, nutrition, and exercise programs, because the tests determine if DNA damage has been reduced and cell health has improved.

Here's a real-life example of how periodic PGH Gene Makeover testing works. Mr. B., a very active businessman and CEO of a major company, had been on the PGH Gene Makeover program for four months. Although he thought he felt better, seemed to have more energy, and was sleeping better, he was not quite convinced that it was due to his new age management program.

Mr. B. decided to repeat the GPF urine test which he took before beginning the program to see if the program was working. He was happily surprised to see that both his DNA damage and free-radical levels had decreased dramatically from the initial base levels taken four months earlier.

But, being an ultimate skeptic, Mr. B. was still not convinced. So he decided to stop his PGH Gene Makeover Program and not take his A.M. and P.M. gene-specific repair tablets (more about them in Chapter 6). He was curious to see what would happen, because he thought perhaps the positive results were "the placebo effect" that he'd read about. He also

GENETIC PROTECTION FACTOR

Your individual genetic protection factor (GPF) is more important than your SPF for your skin or knowing your cholesterol level. GPF is determined by a ratio of your DNA damage rates and free-radical levels. Both of these values are directly related to the damage you suffer at the genetic level. Your GPF value is affected by your lifestyle, genetic inheritance, and the harsh twenty-first century environment. It is your GPF that determines how quickly you age as well as the quality of your health and well-being.

thought perhaps he was just feeling better and he didn't really need this new approach to preventive medicine and optimal aging.

I received an urgent e-mail from Mr. B. about fourteen days after he started his experiment, insisting that I send a new test and a new supply of the PGH Gene Makeover program tablets by overnight FedEx. He said he felt very sluggish, was sleeping poorly, and seemed to be short tempered as well as forgetful. He took the test immediately.

Two weeks later, I received the results of Mr. B.'s repeat GPF urine test. Not surprisingly, it documented values almost as high as his initial DNA damage and free-radical-level test before he began the program. I put this in my file and was curious if I would hear from Mr. B. again. Approximately one week later, I received a new e-mail from Mr. B., stating that he had seen his repeat test values, had gone back on the PGH Gene Makeover program, and was now completely convinced about its effectiveness. I wrote to Mr. B. and said that this is a common experience I've had with many patients over the last few years. They frequently stop the PGH Gene Makeover regimen, for whatever reason. They then revert to a state of suboptimal health. This is usually what makes PGH Gene Makeover program users completely committed to long-term use of their individualized supplements. Only then are they willing to include in their daily regimens other positive lifestyle changes, like exercise and eating correctly. These other changes are essential components for creating the optimal environmental conditions needed to act jointly with and enhance the effects of the PGH Gene Makeover program. I explained to Mr. B., only a combination of both the PGH Gene Makeover program and positive environmental lifestyle changes can maximize the utilization of our genes. (You'll learn about positive environmental lifestyle changes in Chapter 4.)

OPTIMAL AGING VERSUS ACCELERATED AGING

When we look around at people in our family and community, it seems that most people look their age. But some people look years or even decades younger than they really are. The difference in their appearance is directly related to many factors such as lifestyle, diet, exercise, stress, and where they live.

Is positive aging just good genetics combined with a little luck? Until recently, that was pretty much the case. But what we've learned over the last half century by studying people's lifestyle in relation to their health and longevity is that there is a distinctly obvious trend: people who live healthier lifestyles live longer, healthier lives. From a PGH Gene Makeover perspective, the two most important factors are:

1. Good cell function and positive gene expression on the individual level. This means that the genes an individual inherited are being used efficiently at the cellular level.

2. Good copies of cells are made when original cells wear out and are replaced due to youthful repair of damaged DNA and regular apoptosis. That means a person's cells have normal functioning DNA versus poor functioning DNA.

So, luck aside, our new understanding of how lifestyle and environmental factors affect the inside of each cell gives us the opportunity to be more precise in our lifestyle and environmental choices, so we promote positive aging and avoid negative aging.

The Third Secret Revealed: It's Not Just Your Genes— It's Your Total Environment and Lifestyle that Really Matters

*It is by logic that we prove,
but it is by intuition that we discover.*

—HENRI POINCARÉ

The previous chapters discussed the primarily internal factors that have numerous effects on the expression of your genes. When it comes to your general cellular health, however, additional forces of nature are also at play. We group these additional factors together as environmental factors. In fact, the interaction of these environmental factors with your genes has given rise to a new branch of science called *epigenetics*. But before we discuss epigenetics, we need to review some background information about the environmental factors, which are typically grouped in two general categories: *biotic factors* and *abiotic factors*.

BIOTIC AND ABIOTIC FACTORS

When most people think of the environment, what typically comes to mind are things like rocks, soil, water, plants, animals, and air. While these things do indeed make up the environment, there are additional factors we need to consider. To understand what these factors are, divide them into two groups: the biotic factors deal with living things in the environment, and the abiotic factors deal with nonliving things in the environment. This means that plants and animals are biotic factors that have an effect on us, and we have an effect on them, too. For example, most aller-gies are triggered by biologic matter in the air we breathe or by animal bites like bee stings. Abiotic factors like temperature, water and air quali-ty, and sunlight also affect all living things. Human behavior also has an effect on them.

The Biotic or Living Environmental Factors

Examples of biotic factors in the environment that interact with our genes and health include substances in the food we eat, which can exert posi-tive and negative effects. There are the numerous microbes, or *infective biological agents,* that infect us, like bacteria, fungi, and viruses. There are also the friendly microbes that live on and in us, like the probiotic bacte-ria that inhabit our intestines. These bacteria sometimes help ward off harmful microbes or even make vitamins and other healthy substances we need. Plant life in our environment gives off the precious oxygen we need to survive. It doesn't take many examples to realize that the plants and animals around (and in) us do indeed affect us in many ways.

Then there is the food supply we depend on for survival. Until about seventy years ago, the American diet consisted mostly of wholesome whole foods, free from chemicals. But during the past half century, pesti-cides, artificial fertilizers, and other chemicals—initially used to grow more and better crops—entered our food supply. To top all that off, more and more processed foods have appeared on grocery shelves. These processed foods, typically low in wholesome, high-quality nutrition, are instead loaded with too much sodium, too many simple or overprocessed carbohydrates, too many chemical additives, and a lack of essential nutri-ents. So there's been a major change in our food supply—for the conven-ience of creating cheaper foods that have a long shelf-life but are not really nutritious—foods for profit, not for optimum nutrition. Fortunately over the last twenty years, there's slowly but surely been a revival of eat-ing healthy foods, as more people shop at health food stores and eat organic foods, free from chemical additives. There's more about this important topic of nutrition in Appendix A, Gene-Smart PGH Gene Makeover Nutrition.

The Abiotic or Nonliving Environmental Factors

Nonliving or abiotic environmental factors also influence the expression of our genes and the outcome of our health. Abiotic environmental factors include the air we breath and the water we drink; the soil, pumped up with chemicals, that we depend upon to grow the plants we eat or feed to our livestock; and the climate, which includes ultraviolet (UV) sunlight.

In addition, there's a subgroup of abiotic substances called *xenobiotics*. These manufactured chemical compounds, which include dioxins, polychlorides, biphenyls, and other carcinogens, are foreign to living organisms. They make their way into the environment and have both positive and negative effects upon us and other living things.

Light is a major abiotic factor. First and foremost it makes life on planet Earth possible. It is the sun's light energy that plants use to grow, through a process called photosynthesis. However, the direct effects of light on humans can be both good and bad. An example of a positive aging effect is that exposure to light stimulates the production of vitamin D in our bodies. An example of a negative aging effect occurs when we are overexposed to harmful UV rays, which create damaging free radicals in our bodies, leading to skin aging and even skin cancer and death in some people.

The Stress Factor

Researchers are finding that the stress factor, which can be both biotic and abiotic, is very important in influencing our health and gene expression. Stress comes in many forms, but the net effect of bad stress is that it disrupts our metabolism and hormonal balance and produces a state of accelerated or negative aging. Even when all other factors promote health, too much bad stress will have adverse effects on our health at the cellular level.

In addition to stresses of biological origin, abiotic stress factors like climate, chemicals, sunlight, and free radicals affect how our DNA works. The psychological factors that cause stress are so important that Chapter 5 is dedicated to this topic.

ENVIRONMENTAL HEALTH SCIENCES AND EPIGENETICS

The environment's effect on our health is so important that it has lead to the creation of a new branch of medicine called environmental medicine. In fact, the United States' National Institutes of Health created a new institute devoted to this research in 1969 called the National Institute of Environmental Health Sciences. Dr. David A. Schwartz, the director of this institute, summed up the importance this new branch of medicine can have on our health when he said in his inauguration speech: "As a physician-scientist, my ultimate goal is to use environmental sciences to help people live longer and healthier lives." We share a common goal with the mission of this institute: to reduce the burden of human illness and disability by understanding how the environment influences our genes and the development and progression of human disease.

It can no longer be argued whether genetics or environment has the greater impact on our health; the new realization is that they are linked. Epigenetics is a new branch of science that deals with sorting out how both kinds of environmental factors, including lifestyle, affect the expression or function of our genes and gene activity. It deals with how these factors can change the way our body reads our genetic information to make us what we become—in other words, understanding how these factors influence molecular mechanisms at the cellular level, which ultimately controls our gene activity.

Epigenetics is becoming one of the most active areas of scientific research. One area that led to our understanding of how gene/environment interactions affect what we become was the study of identical twins. Researchers all over the world soon discovered that although identical twins have exactly the same genetic information in each of their cells, they often look, function, and behave much differently. This study led to the understanding that being exposed to different environmental factors will result in differences of gene expression.

Case Study of Environmental Factors Affecting Identical Twins

A case study of identical twins is probably the best illustration of how important the environment really is to our health and longevity. In this study, identical twin brothers, who were separated at birth, were reunited after one of them became ill. They accidentally found out about each other when a picture of the one who was a successful Wall Street businessman in New York City happened to be seen by his then unknown brother on a national TV show. The brother, a farmer, had grown up and still lived in a small town in Idaho. To the Idaho brother, the New York businessman, who looked like his clone, appeared to be identical in every way—from his facial features to his idiosyncratic postures. Intrigued, the Idaho farmer sent his photo to the New York businessman, who was also amazed. After a number of phone calls, they decided to meet in person since the Idaho farmer was planning a trip to New York anyway.

The Idaho farmer was shocked at his double. While he thought the man he met was at least twenty years older and obviously quite ill, the New York businessman was equally shocked that the man reflected an image of how he had looked twenty years before. After they traded histories and did some personal detective work, they eventually found out they were identical twins who had been adopted at birth by families who decided to move to two different areas of the country to raise their new boys. The photo of the New York businessman on TV was actually taken eighteen years before.

Why did the Idaho farmer still look like the man in the photo while his New York counterpart looked at least twenty years older? The answer was simple. Although both started out with identical genes, the environment of the inner city, the stress of city and work, a fast food diet, and the lack of fresh air and exercise had a markedly different effect on the genes of the New York City businessman. That environment rapidly accelerated his aging process and deteriorated his health.

These two sets of identical genes placed in diametrically opposed environments illustrate one of the key points we stress in the PGH Gene Makeover program. It is the quality of your genes plus your environment that together dictate your health and your aging process. We need to remember that we inherit genetic potential and not genetic certainties. When we realize that the quality of our health is, to a large degree, placed within our own hands, we can utilize the information that has recently become available to optimize both our health span and life span.

ENVIRONMENT INFLUENCES SNPS

Genetically speaking, human beings are 99.9 percent the same. The only difference between a six-foot, blonde-haired, blue-eyed male and a five-foot, black-haired, brown-eyed female exists in just 0.1 percent of our 30,000 genes. As we learned in Chapter 1, these differences are based on SNPs (single nucleotide polymorphisms, pronounced "snips"), and they occur when a single nucleotide in the gene sequence is altered. To be considered a true SNP, that gene variation must occur in at least 1 percent of the population. While SNPs do not cause disease, the most recent research reveals that they may help determine the likelihood that someone will develop a particular disease or respond in a particular way to a drug, nutrient, or environmental factor.

An even more significant finding has emerged from research into Alzheimer's disease. Scientists now know that environmental factors—notably, lifestyle—play a major role in whether or not a person who is genetically predisposed to Alzheimer's may actually develop the disease. That's an example of epigenetics at work.

One of the genes associated with Alzheimer's, called apolipoprotein E or *ApoE,* demonstrates how SNPs affect disease development. This gene actually contains two SNPs, which results in three possible gene variations, referred to as E2, E3, and E4. Each of these gene varieties, also called alleles, differs by just one DNA base, and the protein product of each gene differs by just one amino acid. Each person inherits one maternal copy of the apolipoprotein and one paternal copy. Here is the profound part: an individual who inherits at least one of the E4 allele gene variations will have a greater chance of developing Alzheimer's disease. While this seems quite incredible, it is believed that the change of just one amino acid in the E4 protein alters the structure and function just enough to make disease development more likely. On the other hand, someone inheriting the E2 allele seems less likely to develop Alzheimer's disease. Apolipoprotein is just one genetic factor linked to Alzheimer's disease, and it is likely that more are involved. But ultimately, above and beyond which allele a person inherits, the critical determining factor in whether a person is afflicted by Alzheimer's is their total environment, especially their diet, nutrition, and exercise. (See Appendixes A and B for more about these topics.)

Research shows that most degenerative diseases, like heart disease, diabetes, type 2 diabetes, or cancer, may be caused by variations in SNPs. The good news is that tests are now available to identify many of the SNPs related to developing these diseases. Because we know that lifestyle plays such a crucial role in triggering SNPs variations, following the PGH Gene Makeover approach means you can take action to change your lifestyle to increase your positive aging outcome.

The tiny bit of variation in our genes makes all the difference in how we look and how we handle the aging process, as well as in our individual needs for specific supplements and nutrient compounds. While this 0.1 percent difference in our genes seems very small, it is this minute difference that determines whether the genes you inherited from your parents are functioning ideally to maintain optimal health or if they are slower than optimal, predisposing you to disease. Now we can test our genes and use the resulting information to our individual advantage—adopting a gene-smart positive aging lifestyle to achieve the ultimate goals of increasing health span and life span.

THE PGH GENE MAKEOVER ENVIRONMENT

We now know that the environment in which we place our genes dictates or determines the outcome of our aging and our health. The Personal Genetic Health environment goes beyond other definitions. It's composed of four essential factors:

1. All the biotic and abiotic environmental factors we have just discussed.

2. Lifestyle and habits. These include our diet, exercise habits, stress levels, and even the supplements or vitamins/minerals we take. These topics will be discussed in later chapters and the appendixes.

3. Belief system. Our beliefs can affect our genes. Our parents were brought up to believe that "if my father had diabetes, I'm sure to get it" or "if my mother had breast cancer, then I'll likely get it." Studies show this kind of negative thought process can become a self-fulfilling prophecy, and can markedly increase your chances of succumbing to diseases of aging. So we need to adopt a new belief system with posi-

tive thoughts for positive aging. For more on the importance of adopting positive beliefs, see Appendix B, PGH Gene Makeover Anti-Aging Exercise.

4. Social conditioning. Each generation is conditioned through television, books, movies, music, and other aspects of culture about what people should act like, look like, and even feel like as they grow older. We incorporate these beliefs into our subconscious and they become part of the factors that control how our genes work.

We've all been led to believe up until now that we were basically victims of our genes, not masters of our fate. (See Figure 4.1.) We now know that is definitely not the case. We have the power and the information to change our future health and longevity. (See Figure 4.2.)

Figure 4.1. Old school of thought about DNA reflects linear thinking, and omits the fact that we can affect our own cellular health.

Figure 4.2. New school of thought about DNA shows that information flows in both directions. The expression of DNA is linked to external and internal environmental signals, which are constantly interacting to turn DNA on and off, creating a dynamic balance. This model reflects the realization that we can affect our cellular health by using the information presented in this book.

ANOTHER FACTOR IN THE PGH GENE MAKEOVER ENVIRONMENT: YOUR HORMONES

Up to now in this chapter we've focused mostly on the external environment. But there are very powerful internal environmental forces that profoundly affect our health and longevity. Hormones are important substances that are part of the intricate environment/gene regulatory control process. The production of our hormones is affected by environmental factors and aging. For example, women going through menopause and men going through andropause are driven by age-related hormonal factors. You often hear about the search for youthful substances like growth hormone and DHEA. Occasionally you see news coverage of how hormones and hormone-like substances get in our food or water. Your body composition and physical activity also affect your hormonal status. For example, higher amounts of body fat cause all sorts of unhealthy hormone imbalances. Good hormonal health is actually a sign of healthy gene expression, which is controlled by your lifestyle and the other environmental factors.

Some of the major hormones that influence your health and longevity include growth hormone, testosterone, estrogen, DHEA, insulin, cortisol, and melatonin. More recently, a new hormone associated with weight control called leptin has entered the media limelight. By following the Personal Genetic Health Program presented in subsequent chapters, you will be able to influence the production and control of these important hormones to improve your health and make the best of your genetic deck of cards.

Hormonal Changes with Aging

Hormones by definition are substances produced by glands in the body. These substances move through the bloodstream to exert particular effects on target tissues and bodily functions. Hormonal changes that occur during aging affect the structure, function, and health of your body in many ways, and they also affect your body composition. Body composition is the amount of body fat in relationship to the muscle, bone, and water in the rest of your body. That portion of your body without fat is often referred to as the fat-free mass or lean-body mass. News stories report that people with high amounts of body fat have higher rates of many age-related diseases such as high blood pressure, stroke, heart attack, hardening of the arteries, type 2 diabetes, arthritis, and cancer.

In addition to being influenced by your lifestyle and environmental factors, body composition has the tendency to change with aging. The trend with aging is for an increase in body fat and a decrease in fat-free mass. While a certain amount of change is inevitable, you can make lifestyle changes to slow down this trend and be healthier for a longer time. One of the common characteristics among centenarians around the world is they have good body composition or lean body mass.

As hormone production and balance falter with aging or from an unhealthy lifestyle, it causes various effects on the status of your body composition and your state of health. For example, a decline in growth hormone and testosterone production decreases fat-free mass and increases body fat. Higher body fat is known to cause problems with insulin function, increased clogging of arteries, and neutralizing the appetite control effects of leptin. (For more about the importance of body composition and lean body mass, see Appendix A, Gene-Smart PGH Gene Makeover Nutrition.)

A QUICK REVIEW OF SOME OF THE MAJOR HORMONES

Here is a quick review of some of the major hormones you will read about in subsequent chapters. Similar to other aspects of good health, maintaining youthful hormone production and healthy hormone balance is vital to longevity and disease prevention. It's essential to improve the levels of the good hormones and reduce the levels of the bad hormones as the body ages. Figure 4.3 summarizes some of the major glands and the hormones they produce.

Hormone Changes With Aging

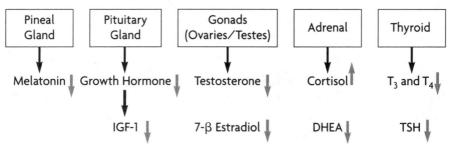

Figure 4.3. Five key glands excrete hormones that undergo changes as the body ages.

- **Cortisol** is a hormone secreted by the adrenal glands that the body needs to break down molecules for metabolism. However, too much cortisol in the body causes excessive damage and accelerates aging. Circulating cortisol levels do change significantly over time, and activity is increased in fat cells, which convert inactive cortisone (the precursor molecule of cortisol) into active cortisol. As fat mass increases with aging, this causes increased production of inflammatory cytokines and stimulates more cortisol production. Stress also increases cortisol levels. Elevated cortisol is related to the breakdown of collagen and elastin in skin (causing wrinkles) and in joints, muscle, and bone (causing pain). Too much cortisol has a negative effect on thymus gland production, while DHEA protects the thymus gland. Along with a host of other health problems associated with high cortisol levels, clinical depression can occur as a result of a decrease in some brain chemicals that is related to elevated cortisol.

- **Dehydroepiandrostenone (DHEA)** is also produced by the adrenal gland. Its maximum concentration is achieved during the third decade of a person's life. From then on, the level of DHEA declines steadily, and in the very old DHEA levels are only 10 to 20 percent of those in young adults.

    DHEA is formed from cholesterol and is the precursor molecule to testosterone in both men and women. Both DHEA and cortisol are produced by the adrenal glands, and their secretion is regulated by a hormone called adrenocorticotropic hormone (ACTH) that is secreted by the hypothalamus. ACTH-induced secretion of DHEA is reduced in the elderly whereas cortisol secretion is increased. DHEA increases immune system response. DHEA deficiency in men causes poor immunity, depression, low metabolic rate, and low good cholesterol (HDL) levels. DHEA deficiency in women can cause decreased libido, poor immunity, depression, and low metabolic rate with weight gain. Studies conducted of both men and women show favorable health benefits from DHEA supplement use, including better mood, more energy, and better body composition.

    In excess, DHEA causes acne, greasy skin, loss of scalp hair, facial hair growth, and high libido. Excessively high DHEA levels in men cause high estrogen levels.

- **Testosterone**, the male hormone, is produced in the gonads (testes) as the result of stimulation by two pituitary hormones. It is also produced in women during the production of estrogen and in very small amounts by the adrenal glands. Folic stimulating hormone (FSH) and luteinizing hormone (LH) are involved in controlling the production of testosterone, which is necessary for the male physique and masculine characteristics like facial hair. Testosterone contributes to the libido, maintaining and building muscles, and to fat metabolism. It is necessary for skin tone, bone health, immunity, and a healthy emotional state. Men with testosterone deficiency may get depressed, lose facial hair and have diminished body hair, and be unable to maintain muscle mass or burn fat properly. Excessive testosterone results in overactive libido, reduced scalp hair, and aggressive behavior. Among lean young men the ratio of testosterone to estrogen is approximately fifty to one. With age this ratio decreases to as low as eight to one. That causes abdominal fat deposits as well as muscle and bone loss, which leads to changes in body composition.

- **Estrogen and Progesterone** are the female hormones produced by the gonads (ovaries) and synthesized from testosterone. Part of estrogen is also produced by the adenoid glands, especially after menopause. The three principal estrogens are estradiol, estriol, and estrone which are necessary for the menstrual cycle, collagen synthesis, brain function and memory, bone mass, and fat cell metabolism. In estrogen deficiency the skin gets wrinkled due to poor collagen tissue structure. Health problems associated with the age-related decline of estrogen during and after menopause are well known. Lower estrogen levels in postmenopausal women are known to impair memory and produce mood swings, hot flashes, and bone loss. An excess of estrodial and estrone has been linked to some overgrowth of ovarian, breast, and uterine tissues and may cause cancer. Estriol has been called "protective estrogen" because it seems to inhibit the excess effects of the other two estrogens.

  Progesterone, the other female sex hormone, counterbalances the water-retention effects of the estrogens and supports the building of new bone tissue. Like estriol, progesterone helps balance the excess effects of estradiol and estrone. It has a calming effect on the brain and reduces the anxiety-producing effects of two of the estrogens. Excess progesterone may be converted to cortisol, which causes rapid aging. Excess progesterone can also cause polycystic ovary syndrome that creates insulin resistance, characterized by obesity, excess facial hair, and no ovulation or menstruation.

- **Growth Hormone (GH or Somatotropin)** is involved in the growth and maintenance of body tissues, including body fat and fat-free body mass balance. Growth hormone production in the pituitary reaches its maximum around twenty years of age. Its concentration in the body is high through the early twenties, and it starts to slowly decline about 14 percent each decade thereafter. Growth hormone has a direct effect on tissues, organs (including muscle, brains, skin, bone), and the immune system. It is converted within the liver to the active form called IGF-1, an insulin-like growth factor (see below). Growth hormone increases anabolic (tissue-building) activity, which builds more muscle, denser bone, and thicker skin. It increases fat burning and improves brain function and the immune system.

  While growth hormone is active and secreted all the time, the majority of adult growth hormone production occurs during sleep and also during and just after exercise. Patients with severe GH deficiency due to pituitary gland damage often suffer from sarcopenia (a degenerative loss of skeletal muscle mass and strength), increased abdominal obesity, and changes in body composition usually observed with aging. It is interesting to note that GH replacement therapy in these patients significantly changes body composition by reducing fat mass and increasing lean body mass. You can maintain youthful levels of GH naturally by taking certain supplements, eating certain foods, and doing the right amount of exercise.

- **IGF-1** (insulin-like growth factor) is a naturally occurring hormonal growth substance that stimulates many processes in the body, including protein synthesis and tissue growth. After activation by the pituitary, IGF-1 is produced by the liver and other cells in the body, including muscle cells. Chemically IGF-1 is considered a peptide (made of amino acids linked together) and gets its name because it is similar in structure to insulin. IGF-1 can bind to insulin receptors located in cells and, in very high amounts, can produce the same effects as insulin. However, most of the effects of IGF-1 are controlled through its own special receptors. Its receptor sites act like a lock-and-key control system where the IGF-1 molecule binds to the receptor site to exert its control functions. (Many other hormones and substances in the body also exert their functions via receptor sites.)

The IGF-1 produced in the liver binds to specific carrier proteins that transport it in the bloodstream to regulate its various biological functions. Beginning in childhood and extending throughout life, IGF-1 plays an important role in growth and development of muscle tissue. So when IGF-1 levels in the body decline with aging, this causes a reduction in muscle mass. It is therefore beneficial to maintain youthful production and levels of IGF-1 for maintenance of muscle mass.

- **Melatonin** is secreted by the pineal gland located in the brain and its secretion is stimulated by darkness. It controls the daily circadian rhythm of sleeping and wakefulness with cortisol. Melatonin is active at nighttime and helps us sleep; cortisol awakens us and is active during daytime. Melatonin is a potent antioxidant and protects DNA from free-radical damage. When not enough melatonin is secreted, we have trouble falling asleep. Breast cancer patients respond better to chemotherapy and radiation therapy when they take melatonin.

Another hormone that was recently discovered is leptin (not in figure), which has revolutionized our understanding of fat tissue. Leptin is secreted by fat cells and signals the brain to regulate appetite and energy levels. It actually tells the brain when the body has had enough to eat. In most people the amount of leptin secreted relates to the amount of their body fat stores, so it helps to balance food intake with body energy stores. However, defects in leptin production and function are known to cause severe obesity. Now we know that fat is not merely a static deposit of stored calories but rather an endocrine organ. Leptin is also involved in other functions in the body such as bone mineral density, the immune system, and the reproductive organs.

THE PGH GENE MAKEOVER AND THE ENVIRONMENT

The importance of the role of both the external and internal environment in regulating genes is integral to Personal Genetic Health. Guess who is in control of making the changes in the environment that can alter how your genes work in a more positive, healthy fashion? *That's right, you are.* The best way to summarize the present state of information about Personal Genetic Health is with the simple equation shown in Figure 4.4. Once you learn your genetic predisposition through personalized gene testing, you will know what to do to alter your environment and push the odds in your favor for better aging and better health from the individual cell level to your complete body.

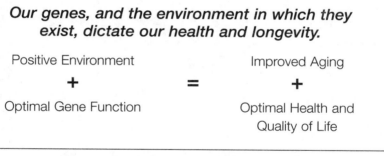

The Complete Picture

Our genes, and the environment in which they exist, dictate our health and longevity.

Positive Environment		Improved Aging
+	**=**	**+**
Optimal Gene Function		Optimal Health and Quality of Life

Figure 4.4. Our genes and the environment in which they exist dictate our health and longevity.

The message of this new equation is profound! You can control your environment, your beliefs, and how you choose to live your life. *You are responsible for your health, not your doctor.* You can now address the reasons for aging poorly and not wait until you are sick to be treated for symptoms while the cause at the cellular level remains untreated or ignored.

You are the first generation to have the power to intercede at the causal level—the level of DNA. This capability is now in your hands, and the first step in taking charge is with a personalized genetic health evaluation.

Information changes our beliefs, beliefs change our habits, habits change our lifestyle, lifestyle changes how our genes work.

So the equation becomes:

Genes
+ environment (profoundly affected by our habits and beliefs)
_____

the quality of our health and longevity

It's important to keep in mind that our genes tell us our odds or tendencies. The environmental factors we introduce into our lives can change or improve our odds and decrease our tendencies for specific age-related disease problems.

5

The Biology of Belief:
What You Believe
Can Make a Difference—
Whether You Believe It or Not

Man's mind, once stretched around a new idea,
never returns to its original shape.

—OLIVER WENDELL HOLMES

In Chapter 4 we noted that the power of our beliefs can change our environment so profoundly that it alters how our genes actually produce proteins and, therefore, create positive and negative effects on every cell in our body. Now we need to look at the effects beliefs can have on two levels of health:

1. The surface or tangible effects of beliefs; that is, how beliefs affect our actions.

2. The subsurface or intangible effects of beliefs, or effects of energy and thoughts, on matter, specifically our DNA.

In Chapter 4 we defined the tangible effects of beliefs with the equation:

Genes + Environment = Positive or negative health and aging + longevity

Our environment is composed of or affected by three interrelated factors:

1. Our beliefs and emotions   2. Stress   3. Lifestyle habits

THE POWER OF BELIEFS

The concept that our beliefs affect our environment and thus profoundly affects our health represents a recent major change in scientific thought. The dogma of the past forty years had been what was called the "primacy of DNA." That is, our DNA and genes control our health, and in effect "we are simply victims of our genes and their genetic programs." In other words, "if it's in your genetic deck of cards, there's not much you can do about it." DNA was viewed as a one-way flow of information, and if we believed and incorporated this dogma into our belief system, we became "victims of our genes."

"New School" Thoughts about Beliefs and DNA

The new paradigm of DNA that has emerged over the last few years is radically different. Instead of the "primacy of DNA," our new thinking centers on the *primacy of the environment*." This new paradigm views the environment as the key factor, which then initiates changes in DNA, which then creates changes in RNA, which then creates changes in proteins, which are the major contributing factors affecting our health and aging process.

The latest genetics research shows that information flows primarily from the environment, which affects DNA, RNA, and proteins, and that the proteins we finally produce are responsible for information or feedback that affects RNA and DNA. So there's a two-way flow of information. (Review Figure 4.2 on page 52.)

This new concept is revolutionary because it states: "To a very large degree you are responsible and in charge of your health and how you age from the expression of your genes on up." What we do with this information is the key! Action is required. So if this new concept makes many of us uncomfortable, just remember: Being uncomfortable is a necessary prelude to change.

Information affects our beliefs. We constantly seek new information via television, Palm Pilots, DVDs, and the Internet, which link us to a constant stream of worldwide information that changes what and how we think. We also seek ways to make this new information make sense. That's why we expand on the new equation:

Genes + environment (affected by information, which affects beliefs, which affect habits, which affect actions that alter our environment) **= positive or negative health and aging**

The essence of this new equation is profound. You can control your environment through your beliefs and the information you decide to use or not use. You are responsible for your health, not your doctor! Now you can address the causes of aging and not wait until you are ill and need to be treated for *symptoms* by an outdated medical model while the *causes* remain untreated or ignored.

Case Study Illustrates the Power of Belief

There is a famous clinical story from the 1950s that illustrates how the power of thinking can indeed reverse genetic diseases. This case study involves a boy whom doctors diagnosed with a bad case of warts, causing his skin to develop a rough, leather-like texture. Dr. Albert Mason decided to try to resolve this boy's terrible skin condition using hypnosis. Dr. Mason reported in the 1952 *British Medical Journal* that when he told the boy under hypnosis that his arm would begin to heal, sure enough, just a week later the boy's arm started improving.

But there is more to this story. When Dr. Mason took the boy to visit a surgeon who had tried unsuccessfully to treat the boy's arm with skin grafting, an important point about the boy's condition was revealed. The surgeon told Dr. Mason that the boy did not have an extreme case of warts. The boy's skin condition was due to a genetic disease called congenital ichthyosiform erythrodermia. Dr. Mason's hypnosis ordered the healing powers of the boy's mind to do what was thought impossible: clear up the boy's genetic skin condition.

Stories about the healing powers of the mind are found throughout the world and throughout history. Our Western medical community is just warming up to the fact that therapies of the mind can be used to promote health in the body. In 2003 Dr. Philip Shenefelt reported in the journal *Dermatologic Therapy* about a research studies report summarizing the benefits of hypnosis, cognitive-behavior methods, and biofeedback on a variety of dermatological diseases. Some examples of diseases reported to respond to mindful healing include acne, herpes simplex, psoriasis, and rosacea. There is also an impressive, growing body of research about how meditation, spirituality, and prayer produce health effects in people afflicted with all types of disorders from anxiety to cancer to high blood pressure.

Emotions: A Focused Form of Thought

In a study from Stanford University discussed by Dr. William Tiller in *Space, Science, and Transformation,* the direct effects on DNA of emotions, which are really focused thought, were studied in human beings. A group of subjects were asked to hold test tubes of water containing human DNA in their hands. They were told to think loving and compassionate thoughts by visualizing someone they loved, thus triggering positive emotion. The DNA in the test tubes was meticulously measured and shown to actually and clinically change shape: the DNA uncoiled. That is the exact same movement that occurs when genes are activated. The regulatory proteins around DNA relax and open, thereby allowing the genes to be turned on.

When this same group of subjects was told to focus on a negative emotion by thinking of someone they disliked or hated, the DNA was again measured and found to coil tighter. This created a response similar to the state when genes are restricted and their activation inhibited.

A follow-up study by Dr. Tiller showed the exact same effects even when the test tubes were not held by the subjects but were place in racks on the other side of the room, and positive emotions were focused on the test tubes more than ten feet away. This is evidence that emotions of others, as well as ourselves, may affect our health!

Although such evidence is not yet fully accepted by medical science because of the magnitude of the implications, many open-minded researchers believe there is no question that our thoughts and emotions interact with our genes and affect our health and perhaps even the health of those around us.

Beliefs Are Energy

Our beliefs are not just thoughts; they are in essence electromagnetic fields (EMFs) based on emotion and information. A study published over forty years ago compared the efficiency of information transfer between energy signals (EMFs) and chemical signals (molecules) from biological systems. It revealed that EMF signal frequencies are 100 times more effective in transferring information than such physical signals as hormones, neurotransmitters, or growth factors.

Recently research has shown that EMF information can even be recorded and e-mailed over the Internet to a laboratory halfway around the world, where it was played back to cells that responded as if the physical molecules were actually given to the cells! In the world of cell biology, EMFs are now being used to turn on or turn off different genes to produce changes in proteins similar to those produced by medical drugs—without the medicine. Clearly the present model of medicine and drugs needs to be updated to incorporate these new findings.

As we now know, our beliefs influence the most intimate level of our biology—our DNA. If we fill our life with positive thoughts and emotions about our health and our personal power to affect how we feel and how we age, we can change the quality of our life as well as our destiny. We are the first generation of human beings who have the power and information to intercede at the level of cause—at the level of DNA.

STRESS

Stress adversely affects all of us; some more than others. It comes in many forms—from our own thoughts, from the actions of others, from work and everyday life, and from chemicals in our food, water, and air. Understanding some fundamentals about the stress/health connection and taking actions to control stress in your life is an important part of the Personal Genetic Health approach.

What Is Stress?

Stress is a normal part of life, and a certain amount is essential for survival. It allows us to react appropriately to either real or perceived danger through the flight-or-fight response, which triggers a quick elevation of insulin and cortisol (the two stress hormones), allowing us to flee a potentially dangerous situation. Stress also helps us mount a defense against bacteria or viruses. An imagined threat or recall of a dangerous event provokes a stress response containing all the same elements as acute stress from trauma or infection. Illustrating this point, researchers at the National Institutes of Health have found a connection between depression in adults and traumas experienced in early childhood.

The primary sources of stress in modern society are time pressures due to an overcrowded schedule, constant stimuli, and frequent interruption of thoughts or repetitive thoughts. We are often required to handle multiple activities simultaneously. When we multitask we are denied the opportunity to be in the present moment, which leads to subtle but chronic changes in hormone levels, primarily in cortisol.

We are also frequently exposed to various stressors, such as rudeness, anger, anxiety, restlessness, noise, crowding, isolation, hunger, danger, emotional upheaval, psychological trauma, and infection. Real harm can come from something as routine as commuting to work and exposure to infections from those around us. Most of us are aware of these daily stressors, but we are less attuned to environmental stressors.

Our modern world exposes us to radiation from electromagnetic fields, surgical devices, radiological sources (x-rays), and toxic elements in water, food, air, household cleaners, exhaust fumes, pesticides, and sprays. Some of the most frequently encountered electromagnetic stressors are from hair dryers, cell phones, cordless phones, electric blankets, and wireless communication devices. Computer monitors emit magnetic radiation from the back, so you should sit at least three feet away from them.

Though cell phones are ubiquitous in modern society because they offer us a way to keep in close contact with colleagues and family members, we now know that even regular cell phone use may be very hazardous. A landmark study by Leif Salford, M.D. of the University of Lund, Sweden, published in *Environmental Health Perspectives,* found that adolescent rats exposed two hours a day for fifty days to a device emitting one-sixth the radiation of a digital cell phone experienced either 2 percent brain cell death or 2 percent of their brain cells were in the process of dying. Dr. Salford suggests keeping calls short or using a hands-free device to reduce radiation exposure.

Modern life with its unremitting psychological stress imposes artificial fight-or-flight responses. Finnish researchers published a thirty-year study that followed 812 men and women who worked for a company in the metal industry. The scientists found that those who had the most job-related stress were over twice as likely to die of a heart attack as those in low-stress jobs. Those who felt undervalued for their contribution (low salary, lack of social approval, limited job opportunities) were two and a half times more likely to have a fatal heart attack.

While both a high-stress job and being undervalued increased the risk of death from a heart attack, an interesting difference in coping mechanisms emerged from these two groups of workers. Those who had stressful jobs had high cholesterol while those who had low job satisfaction gained more weight. Both of these conditions stem from elevated cortisol. However, in the first case it caused metabolic imbalance, while in the second it caused insulin imbalance and possibly brain chemical disturbances that led to overeating.

Life Energy, Stress, and Hormones

Your body contains trillions of cells, each a microcosm of molecules vibrating with a specific frequency of energy. They buzz, they bump into one another, and they interact, exchanging energy as they do so and as they drive cellular metabolism. When you are healthy, the energy transfer is smooth and harmonious. Your whole body resonates with the dynamic flow of energy known as *vitality* or *life force.*

The flow of life's energy has been recognized for thousands of years. The ancient Chinese called it *chi* or *qi* (pronounced chee), while the Hindus called it *prana.* More recently, Sigmund Freud called it *libido,* and Henri Bergson referred to it as *elan vitale.* Most Westerners think our life force is confined to the energy within our bodies, but traditionally it has been recognized that we interact with our external environment and energy sources outside our bodies. Consequently, we can draw energy and vitality from our environment or shut it out. It's a choice we exercise every day.

Low energy and ill health have been viewed in traditional medicine as unbalanced energy within the internal and external environment of the body. Healing has focused on removing blockages to the flow of energy. How does energy transfer occur on the cellular level and what effect does stress have on it? The answer is related to the hormone cortisol.

We discussed several hormones including the sex hormones in Chapter 4. Hormone release in the body is a carefully controlled step-by-step process. In this chapter we are interested in the effect of stress on the hormone cortisol.

Stress activates a part of the brain known as the hypothalamic-pituitary-adrenal (HPA) axis, and it releases peptides including corticotropin-releasing factor or CRF. These peptides serve as the central coordinators for neuroendocrine (hormone), immune, and behavioral responses to stress. CRF is specifically concerned with the adrenal glands, and it stimulates the adrenal-activating peptide adrenocorticotropic hormone or ACTH to release the hormones cortisol and glucagon. These hormones address the stress response by altering behavior, triggering release of multiple peptides, and changing involuntary nervous system functions such as digestion. Cortisol and glucagon also change the way glucose and oxygen are utilized by the brain and the muscles needed to get you out of danger.

In addition to cortisol and glucagon, the adrenal glands secrete the neurotransmitters epinephrine (adrenaline) and norepinephrine (noradrenaline). These two neurotransmitters stimulate the brain and activate the fight-or-flight response, with which we are all too familiar. The effect of all this molecular exchange is a series of lifesaving events that protect you from infection or injury. However, chronic stress puts the body on constant high alert, depleting life energy needed to promote health and longevity.

Glucagon speeds up the conversion of muscle tissue to glucose to provide a ready source of fuel for molecules engaged in the stress battle. A glucose surplus eventually leads to weight gain and inappropriate insulin response. During stress-free periods, glucagon is balanced by insulin, which keeps blood sugar under control and promotes uptake of blood sugar into cells and conversion of proteins into muscles. However, during stressful periods, the glucagon/insulin balance is tipped and protein breakdown (catabolism) is increased. High levels of cortisol enhance this effect. By far the worst effect of high cortisol is accelerated aging.

Cortisol and Aging

Cortisol affects every cell of your body and your central nervous system, activating protective defense mechanisms to save your life. However, per-

sistent elevation of cortisol disrupts homeostasis by interfering with hormone, immune, brain, and nerve function. For example, high-stress states
produce lapses in short-term memory, affect appetite, and weaken your
immune response. Research by Stanford professor Dr. Robert Sapolsky has
shown marked brain cell damage as well as immune cell deterioration
from chronic cortisol elevation. You know how easy it is to get a cold or
other ailment when you've been highly stressed.

In an amazing feedback loop, rising levels of certain neurotransmitters, hormones, and peptides alert the brain that the stress is still there,
and the adrenals kick in more cortisol. This is extremely useful when we
are threatened with bodily harm or attacked by microbes, but it's of little
use when we're stuck in traffic on the freeway. Eventually, high cortisol
levels deplete the adrenal glands of cortisol reserves and initiate an
unremitting stress cycle that ultimately leads to adrenal exhaustion and
chronic disease development.

Cortisol has been dubbed the age-accelerating hormone. The main
function of cortisol is to reduce inflammation in case of injury, and the
drug cortisone is used for that purpose. However, as we age, cortisol levels rise. The more stressful our lifestyle and the poorer our diet, the higher the cortisol levels. High cortisol levels are directly related to the
following major factors in aging:

- Breakdown of collagen and elastin tissue in skin, joint, bone, and muscle tissue.

- Nervous system damage.

- Memory loss and decreased cognitive function.

- Decreased immune function, leading to increased susceptibility to
  infectious diseases.

- Increase in inflammatory substances.

- Fat metabolism disorders reflected in elevated triglyceride, total cholesterol, low good-to-bad cholesterol ratio, and obesity.

- Body fluid retention and high blood pressure.

- Decreased hormone function.

- Increased sugar cravings due to increased insulin levels.

- Increased inflammation due to allergies, asthma, and arthritis.

- Skin problems, including wrinkles, acne, psoriasis, seborrhea, and alopecia (hair loss).

Increased cortisol alters the function of other hormones, including growth hormone, insulin, thyroid-stimulating hormone, and sex hormones. Neurotransmitters and neuropeptides are also affected. The immune system receives these chemical signals and communicates messages back to the central nervous system, which involves small protein substances called cytokines that directly affect the brain. This communication alters immune response, inflammation, and pain perception. That's why constant stress produces illness, most notably head and body aches. As noted in the list above, stress and elevated cortisol levels also cause such skin disorders as acne, seborrhea, hair loss, skin aging, and wrinkles. Moreover, stress contributes to reproductive problems, including premenstrual syndrome (PMS), menopausal symptoms, low libido, and lack of fertility.

Constant stress also produces burgeoning numbers of free radicals from your mitochondria as they amp up production of energy to support the molecules involved in the stress response. Your body, especially if it is not well nourished, cannot neutralize the massive numbers of radicals produced during high-stress periods.

Cytokines are important mediators of inflammation. Some increase inflammation while others inhibit it. They are also involved in protein breakdown (catabolism). Like the stress hormones, certain cytokines, especially TNFα (tumor necrosis factor-alpha, which promotes tumor growth and inhibits DNA repair), are associated with loss of muscle protein and increased synthesis of inflammatory proteins such as C-reactive protein. Stress ages you by damaging your proteins, carbohydrates, lipids, and genes.

Stress increases cortisol, inflammatory substances, and free radicals, and decreases sex hormones. All these changes increase production of the group of cytokines that cause inflammation. Low-grade inflammation is the basis of all chronic diseases, including heart disease, type 2 diabetes, dementia, and accelerated aging.

LIFESTYLE CHANGES: NEW HABITS TO CONTROL STRESS

The best way to protect the life of your genes is with stress-control methods. You can choose what works best for you and fits within your personal belief system. Now that you understand that your cells communicate in a harmonious melody on both a physical and an energetic level, you can help orchestrate the music by sending them healthy messages from your emotions.

Be Aware of a New Day

Spend a few extra minutes after awakening to become aware of your body. This means checking to see how you feel and being grateful for what your cells were doing during the night. As you slept, they were storing up energy and busily repairing damage. Hormone levels follow twelve-hour patterns, as do levels of peptides and enzymes. Be grateful for the warmth and comfort of your bed and for the vital state of your health. Welcome the new day and think about the positive things that will happen to improve your well-being.

If you have chronic pain when you wake in the morning, you can use this time to send messages to the pain centers of your brain to increase your natural opiate-like painkilling endorphins. Work to create a positive image of the health you are building and how you will overcome disability. Take several deep breaths, stretch your limbs, and then roll out of bed.

Breathe Deeply and Slowly

Ancient and modern healing arts share a common practice of correct breathing. Meditation, yoga, qi gong, tai chi, and Pilates teach you the importance of drawing air into your lower dantian (qi gong), third chakra (yoga), or core (Pilates). This area behind your navel is considered the center of energy or fire.

Accessing it requires tightening the lower abdominal muscles and drawing air downward toward your lower spine. As you do this, you will be using the back of your rib cage to fill your lungs with air. Most people inhale deeply by pushing the rib cage out toward the front of their body. This increases tension on the shoulders. With lower abdominal breathing, the shoulders naturally drop into neutral position.

Inhale deeply and slowly through your nose and exhale by slowly blowing air out through your slightly open mouth. Purse your lips slightly as if you were blowing the puffy white seeds from a dandelion. Do this deep breathing exercise frequently throughout the day. Visualize white, clean clouds entering your nose and gray clouds leaving your mouth. Do this deeply three or four times. On your last breath, see the white cloud both entering your nose and leaving your mouth.

Meditate Every Day

What is meditation? It is simply learning to extend the time interval between one thought and the next. The longer the gap, the longer the meditation. Meditation has been practiced and perfected over centuries as a process for healing and restoring mind/body. All ancient healing methods used meditation as a way to connect with one's inner being and activate the powerful healing ability of the body.

Meditation was developed by traditional healers within the context of prayer, whereas in holistic medicine it is employed as a healing technique independent of one's spiritual and cultural beliefs. Modern science has now validated the practice of meditation as the way to connect with the molecules of our emotions. Scientists have found that meditation changes the frequency of brain waves and allows us to access other levels of awareness. For example, when Buddhist monks are meditating and cloistered priests or nuns are at prayer, electrical waves in their brains' awareness center (posterior superior parietal lobe) show strikingly low brain wave activity. Meditation usually increases alpha and theta waves which help to decrease cortisol. Research by W. C. Bushell published in *The Annals of the New York Academy of Science* in 2005 suggests that meditation along with dietary restriction may prolong health span in humans. Another study, which Dr. Giampapa performed at the Monroe Institute in Faber, Virginia, shows meditation increases key hormones like DHEA and melatonin as well as decreases inflammation.

Slowing the mind down promotes a feeling of peace and oneness with the universe, which can provide powerful medicine in today's hectic world. To get the best results from meditation, we suggest taking a few classes. These are commonly offered by health clubs, health professionals, and destination centers like the Monroe Institute in Faber, Virginia. Once you are familiar with the basic techniques, then find or create a comfortable place of solitude in which to meditate on a daily basis.

Stretch Every Day

After meditating, spend ten minutes stretching. There are many books on how to stretch, including those on workouts, yoga, tai chi, and qi gong. Some are listed at the end of Appendix B, PGH Gene Makeover Anti-Aging Exercise, and in the Resources in the back of the book. If you exercise in the morning, stretch just before you exercise. You'll find the breathing technique you learned in this chapter will be useful during exercise as well.

Prepare for Your Day

Take time to feel good about yourself before planning your day. The rest of your morning routine should flow from the excellent start you have made. Let your mind, hormones, and immune system establish connections that will buffer you against the day's stressful events. As you move into the next few activities, you'll find your creative mind goes to work. Many people find that solutions to problems or creative ideas suddenly emerge while they are in the shower. Exercise control over these ideas. Let them mature as you eat breakfast and head out the door. Use commuting time to listen to music or inspirational tapes or to read on the train. Avoid using your cell phone, especially when driving. As you move through the other events in your day, always work to keep stress under control.

Relax

Relaxation—the process of calming yourself and making yourself free from stress and tension—is another anti-aging technique that accesses the healing ability of your cells. From our experience, deep relaxation can be dramatically effective in relieving various aging symptoms such as inflammation, anxiety, and muscle tension. Relaxation can be a lifesaver when dealing with stress and life crises. By relieving tension, deep relaxation improves circulation and hormonal balance and accelerates the healing process.

Learning to relax involves learning how to be present in the moment—in the "now"—and how to stop the incessant stream of mind chatter much as one does during meditation. *The Power of Now* by Eckcart Tolle is an excellent source of techniques to learn how to stop this constant flow and escape the pressures of time urgency so endemic to life in the twenty-first century. (See Figure 5.1 below.)

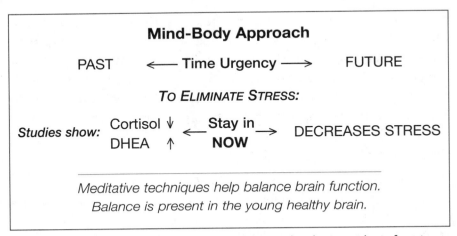

Figure 5.1. Meditative techniques help quiet the mind and improve brain function.

Plan to Beat Daily Stress

You know you will be exposed to stressful events each day. So have a plan to deal with that. For example, taking frequent breaks throughout the day for stress relief through breathing, stretching, relaxation, and meditation will extend your healthy years and add to your quality of life. You need to make your daily grind less stressful by using these anti-stress techniques. When you are about to enter a stressful situation, take deep breaths and continue to do so. It is really that easy. In fact, you will discover that the less stressed you are, the less stressed the people around you will be.

A HOLISTIC VIEW OF HEALTH

What you believe, think, and feel affects your health, your body shape and composition, and your moods and emotions. All of this has a profound affect on your genes. Medicine in the twenty-first century is finally embracing what the ancient healers knew—that we are more than physical beings interacting in a physical world.

6

Personal Genetic Health Nutraceutical Supplements: Forget Daily Once-a-Days—21st Century Anti-Aging Nutrition Is Here Now!

If I knew I would live as long as I have,
I would have taken better care of myself.

—GEORGE BURNS, COMEDIAN, AT AGE 101

n Chapter 1 we discussed the massive confusion surrounding supplements as well as anti-aging programs. Indeed, confusion reigns, and getting good advice about taking supplements is hard to come by. Millions of people are indoctrinated through the media to take "once-a-days" to stay healthy. But can they help you age optimally? The answer is they really can't.

NUTRITIONAL BUILDING BLOCKS

We eat to supply the building blocks that make new cells to replace worn out or damaged cells. The carbohydrates we eat supply energy or fuel for our cells, and protein supplies the backbone molecules for hormones, enzymes, and the amino acids that help build DNA. Fats supply basic elements for cell membranes and cell receptors. Along with these macronutrients, we need the micronutrients that come from vitamins, minerals, herbal extracts, enzymes, and cellular repair compounds. *The quantity and type of vitamins and minerals we take can increase or decrease our gene activity, thereby increasing or decreasing cell aging.* We may need increasing amounts of specific nutrients as we grow older to maintain optimal DNA function and cell repair.

These essential micronutrients are responsible for interacting with our genes and regulating how efficiently they work. They can turn on gene activity or turn down gene activity through direct or indirect mechanisms. As we discussed in Chapter 4, a whole new science of gene manipulation called *epigenetics* has arisen in the past few years. Epigenetics involves the study and understanding of compounds that regulate how efficiently our genes work.

All of us have eating habits that rob us of essential natural nutrients. In general the foods we eat contain excessive amounts of unhealthy refined carbohydrates and saturated fats. More important, we often eat the wrong type of carbohydrates and fats. We are exposed to artificial preservatives, hormones, and chemicals that inhibit our cells and DNA from working at optimal levels and actually damage our cells on a daily basis. In fact, these negative environmental factors interact with our DNA and cause it to malfunction, so that each cell ages much more rapidly than it should and makes poor copies of itself.

RDI Levels Don't Go Far Enough

The Institute of Medicine in the National Academy of Sciences reviews volumes of scientific evidence before making recommendations for all types of dietary intakes. The RDI (Recommended Daily Intake, formerly called RDA—Recommended Dietary Allowance), which you'll find in the nutritional fact panels on food and supplement labels, is a guideline for healthy people and is not necessarily absolute. The expert panels set up to review micronutrient requirements recently established optimum levels and upper limits for micronutrients. However, this is still based on the average healthy person eating an adequate diet, and does not necessarily allow for anti-aging benefits. In fact, it has only recently been established within the scientific community that diet alone may not supply all the micronutrients needed to maintain good health.

In addition, there are many compounds in foods—things like preservatives, fertilizers, and chemical byproducts of cooking—that have been shown to damage cells and cause DNA damage. So we need supplements that take these factors into account, not just the minimum intake needed to prevent nutrient disease deficiencies. But even the lack of disease is not an indication of prime health. We need supplement intake that will result in optimal, positive aging at the cellular level, based on our personal genetic needs. This is supported by the results of a significant clinical study of genetic variations and nutritional requirements reported in 2006 in the *American Journal of Clinical Nutrition.* (See Figure 6.1.)

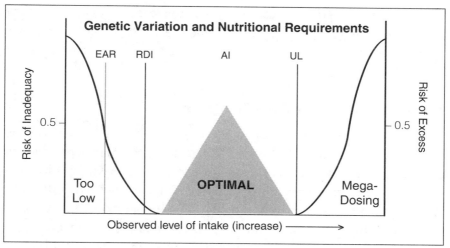

Figure 6.1. A significant clinical study published in 2006 showed that genetic variations could be optimally adjusted using doses of nutritional supplements higher than the RDI.

Why We Need Key Supplements

Once we arrive at approximately age 30, our digestive tract begins to lose its ability to digest food at an optimal level. Additionally, an older person can also have reduced nutrient absorption. Beyond the stomach, organs like the liver play important roles in how nutrients are used in the body. When liver function also starts to decline with age, a decline in digestion, absorption, and utilization of nutrients from the foods we eat also occurs.

Whether from aging or disease, we become unable to extract those essential vitamins, minerals, cofactors, and even carbohydrates, proteins, and fats that we need from the food we eat. The classic example of this is vitamin B_{12} deficiency, caused by the reduction of a digestive enzyme critical in the absorption and utilization of this most essential vitamin. With aging, we are in essence starving ourselves and our genes of key compounds they need to work at their best to keep DNA repair and cell copies at their highest levels of perfection. Therefore, if we don't supplement our diets correctly, we don't allow our genes to work at their full potential and our DNA to repair the damage it suffers, especially in the harsh environment we now find ourselves in. Dr. Bruce Ames, an eminent nutritional researcher, stated in an article published in 2002 that the damage to our DNA caused by the lack of essential vitamins and nutrients is exactly the same as damage caused by radiation exposure. No one in their right mind would purposefully expose themselves to radiation; yet if we consciously and daily avoid the use of key nutraceuticals, we produce the same effect.

Nutrients Can Affect Gene Function

The Personal Genetic Health's new model of nutrition goes well beyond the RDI way of thinking. With the RDI approach, the minimum intake of essential nutrient intake is the goal—just enough to prevent or reduce the rate of nutrient deficiency diseases at the national level. While this RDI approach had some merit in the 1940s when modern nutrition research began, it is obviously outdated now, as it does not even address the concept of optimum daily intake of essential nutrients. Recent research has revealed that essential nutrients are needed in amounts that not only prevent nutrient deficiency diseases, but also make the body work better and improve health.

Our most recent understanding of the interaction of nutrients and gene expression has led to the PGH Gene Makeover's new model of essential nutrition. The nutrients you ingest from foods and supplements can have direct effects on gene expression and activity in three ways:

1. Directly turn off or on a gene's activity.

2. Change the enzyme activity involved in the metabolic pathways the gene regulates, turning their function up or down.

3. Interact with the end product that a gene makes, that is, a protein, making the gene's function more efficient or less efficient.

When we consider the newly discovered functions of nutrients, it's easy to understand why a new approach to supplement taking is in order—one that integrates our knowledge of the roles of nutrients with the newest information about how nutrients can control gene expression, function, and metabolism to benefit our health and longevity.

Supplements Needed to Promote the Five Key Anti-Aging Processes

Each cell needs nutrients and supplements to help regulate the genes and the metabolic pathways that affect the five key processes profoundly affecting cell aging. Studies also show that these same five processes are positively affected by calorie restriction. To review, these vital five key processes are:

1. **Methylation:** The process of adding a methyl group to the chromosomes regulates and actually changes how actively a gene may be working or, even more important, whether genes are "turned on" or "turned off." Genetic abnormalities of methylation have been linked to cardiovascular diseases, elevated homocysteine levels, cancer, and faulty DNA repair.

2. **Inflammation:** The elevation of specific inflammatory compounds both inside and outside the cell interfere with normal cell mechanisms. We see this in patients with joint and bone pain, cardiovascular disease, premature brain aging, Alzheimer's disease, and memory loss, as well as general age-related arthritis.

3. **Glycation:** Glycation is the regulation of blood sugar levels and body fat levels. If faulty genes disrupt this vital function proteins, enzymes, and hormones can become coated, making them inefficient or nonfunctional. This is what happens with diabetes.

4. **Oxidation:** Each cell normally produces its own intrinsic antioxidants, that is, the antioxidants we make naturally that do not come from food. These include superoxide dismutase, catalase, and glutathione peroxidase. These three key intrinsic antioxidants are made in the cell energy organelle, the mitochondria, and they are the first line of defense against free radicals in our bodies. Free radicals are involved in tumor and cancer development and aging because they damage DNA.

5. **DNA repair:** Probably the most key factor in aging is the level of ability to repair the damage to the inherited genetic programs that reside in our DNA. The constant bombardment of our DNA by free radicals as well as damage from preservatives and chemicals which intertwine within the double helix of DNA (called "DNA addicts") can wreak havoc with the internal programs that keep our cells aging in a healthy way. Abnormal DNA repair rates are seen in many diseases, including decreased immune function, cancer, and premature aging syndromes.

From a nutritional supplement standpoint, the good news is that since we are genetically 99.9 percent the same, we all have basic or core nutritional needs. Therefore, at the core of an optimal age management program, we all should be taking a supplement program with essential compounds that help our genes improve these five cellular processes.

Though the only difference between each of us is 0.1 percent of our genes, this small amount makes all the difference in the real success of our personal PGH Gene Makeover programs. Herein lies the answer to what we need as individuals to make our genes, or SNPs, work to their full potential. We can now test our genetic deck of cards and put the odds in our favor to have a healthier quality of life.

Presently, personal genetic health testing is the only way you can know what *your* gene variants are, and what additional supplements *you need,* as opposed to someone else. The test is the foundation for establishing a personalized genetic health and nutrition program. With repeat testing every four to six months, your program can be adjusted as needed.

When our personal genetic health program is working, we see a steady drop or at least a reduction in DNA damage rates and in free-radical levels, the main cause of DNA damage. It works because each of the five key cellular processes are interrelated and affect each other in a positive way.

Of course, supplements alone cannot compensate for a poor lifestyle of frequent fast foods, smoking, lack of exercise, and a stressful job. What's equally important is adopting a reasonable lifestyle and changing your belief system to incorporate the concept that you can make a difference and that, ultimately, you're the one in charge.

Micronutrient Deficiency and DNA Damage

Approximately forty micronutrients are essential for normal metabolism. According to Dr. Bruce Ames, world authority on nutrition and DNA research, "Micronutrient deficiency can mimic radiation (or chemicals) in damaging DNA by causing single- and double-strand breaks, oxidative lesions, or both." These are the root causes of most diseases. Table 6.1 on page 82 lists the most common vitamin and mineral deficiencies, food sources of these vitamins and minerals, and the kind of DNA damage that occurs when they are deficient.

Correcting micronutrient deficiencies requires taking doses higher than the recommended daily intake to activate sluggish metabolic enzymes. Dr. Ames has estimated that about fifty human genetic diseases are due to defective enzymes, and that supplying the appropriate coenzyme and/or cofactor in optimum doses can induce greater enzyme activity.

In addition to preventing or correcting DNA damage, other micronutrients are needed to activate enzymes involved in important metabolic pathways derailed in the aging process. Yet most multivitamin supplements, even those labeled as anti-aging, are missing many of these key micronutrients and provide only the minimum RDI of vitamins and minerals needed to prevent aging. Of course, the popular once-a-days fall radically short of supplying these essential compounds.

Epidemiologists at Harvard Medical School have identified several metabolic disorders that stem from suboptimal intake of vitamins and minerals. Yet these disorders do not fit the classic definition of vitamin deficiency, and standard blood tests do not reveal such conditions. Consequently, physicians are reluctant to routinely recommend vitamin and mineral supplements as disease preventives. The Harvard scientists stress that supplements are not a substitute for a good diet with a wide diversity of phytonutrients, but supplements should be used as secondary therapy in treating metabolic disorders. Given the state of the art, our anti-aging supplement plan recommends nutrients aimed at specific gene groups to correct aging conditions at both the cellular level and the age-related disease level.

Here's a real-life example of why that's important. Mrs. H., a forty-eight-year-old professional businesswoman and a part-time smoker, had

Table 6.1. DNA Damage and Conditions Stemming from Micronutrient Deficiencies

Micro-nutrient: Folic acid   **Food Sources:** Dark green veggies

% People Deficient: 10%

Anti-Aging Daily Intake: 400–1,000 mcg

DNA Damage from Deficiency: Chromosome breaks, base substitution. Required for cytosine and thymine; parts of the DNA molecule.

Conditions Stemming from Deficiency: Colon cancer, cardiovascular disease, brain dysfunction, birth defects

Micro-nutrient: Vitamin B_{12} **Food Sources:** Meat

% People Deficient: 4% get less than half the RDI

Anti-Aging Daily Intake: 400–600 mcg

DNA Damage from Deficiency: Same as folic acid. Required for all four bases making up the DNA molecule: adenine, guanine, cytosine, and thymine.

Conditions Stemming from Deficiency: Nerve damage, breast cancer

Micro-nutrient: Vitamin B_6 **Food Sources:** Whole-grain bread, cereals

% People Deficient: 10% get less than half the RDI

Anti-Aging Daily Intake: 50–100 mg

DNA Damage from Deficiency: Same as folic acid. Required for thymine synthesis.

Conditions Stemming from Deficiency: Same as folic acid and B_{12}

Micro-nutrient: Vitamin C **Food Sources:** Fruits, vegetables, liver

% People Deficient: 15% get less than half the RDI

Anti-Aging Daily Intake: 750–1,000 mg

DNA Damage from Deficiency: Free radical (DNA oxidation) damage is similar to that from radiation

Conditions Stemming from Deficiency: Cataract risk increased 4 times, cancer, cardiovascular disease

Micro-nutrient: Vitamin E **Food Sources:** Nuts, vegetable oils

% People Deficient: 20% get less than half the RDI

Anti-Aging Daily Intake: 200–800 IU

DNA Damage from Deficiency: Radiation mimic (DNA oxidation); damage is similar to radiation

Conditions Stemming from Deficiency: Cancer: colon, cardiovascular disease, immune dysfunction

Micro-nutrient: Iron   **Food Sources:** Meat, fortified cereals

% People Deficient: 7% get less than half the RDI

Anti-Aging Daily Intake: 9–18 mg

DNA Damage from Deficiency: DNA breaks, radiation mimic

Conditions Stemming from Deficiency: Brain dysfunction, immune dysfunction, cancer

Micro-nutrient: Zinc   **Food Sources:** Meat, whole grains

% People Deficient: 18% get less than half the RDI

Anti-Aging Daily Intake: 10–25 mg

DNA Damage from Deficiency: Chromosome breaks, radiation mimic

Conditions Stemming from Deficiency: Brain dysfunction, immune dysfunction, cancer

Micro-nutrient: Niacin   **Food Sources:** Meat, legumes

% People Deficient: 2% get less than half the RDI

Anti-Aging Daily Intake: 100–500 mg

DNA Damage from Deficiency: Disables DNA repair enzymes (polyADP-ribose)

Conditions Stemming from Deficiency: Neurological symptoms, memory loss

Note: Statistics based on 1 percent of the U.S. population = 2.7 million people in 1998. For percentages of the population listed for each micronutrient, with the exception of folic acid, intakes were less than half the recommended daily intake (RDI). Additional micronutrient deficiencies will likely be recognized as time passes.

Source: Adapted from Ames, B. *Toxicology Letters* 1998;Vol. l02–103:-18. Table from Vincent Giampapa, Ronald Pero, and Marcial Zimmerman, *The Anti-Aging Solution,* Hoboken, N.J.: John Wiley & Sons, Inc., 2004.

Refer to the glossary for information regarding DNA and genetic related terminology.

built her own company and had become extremely successful both financially and professionally over the past fifteen years. Highly intelligent and extremely inquisitive, she made a hobby of avidly reading about the latest use of vitamins and minerals as well as about health and longevity. She had been taking once-a-days in the morning as well as a number of other supplements suggested by a number of magazines as well as her vitamin store representative. Mrs. H. was excited to take the PGH Gene Makeover urine test to discover her genetic protection factors (GPF) and find out how successful her personal little experiment had been. (The test and genetic protection factors will be discussed in Chapter 7.) Her results showed excessively high DNA damage rates as well as excessive free-radi-cal levels.

Mrs. H. was quite shocked. After all, she'd taken everything she'd read about in life extension journals and magazines. "How could it be?" she thought. She wrote me an e-mail and asked for help.

What I concluded after studying her test was that Mrs. H. had created a complex condition from excessive unbalanced antioxidant use called oxidative stress. This so-called "abuse" of antioxidants, along with her smoking and fast-food diet and stress on the job had created a situation where she was actually accelerating her aging process. I explained to her that she should be taking more balanced antioxidant combinations along with other categories of supplements or nutraceuticals based on her personal genetic needs.

After cutting her supplements down to just six tablets a day and using the personalized genetic approach, Mrs. H.'s repeat urine test for GPF factors documented a marked improvement. Her DNA damage rates and free-radical levels virtually fell by half within a month. She also felt less stressed, was sleeping better, and was more mentally focused throughout the day than she had been in the past three years. She again e-mailed me to mention that her cholesterol levels had also dropped, and after four months, she let me know that her skin as well as her body composition had never been so good.

Mrs. H. attributed these many changes to the PGH Gene Makeover supplement program as well as to the fact that now she had more energy and had begun to exercise as I had suggested. She was receiving positive feedback from her friends and employees at work, which also helped her continue to maintain the positive lifestyle changes she had incorporated into her personal environment.

For more about taking supplements, see "Answers to Common Questions about Aging and Supplement Use" on page 85.

Quality of Your Supplements Really Matters

Not just any supplements will prevent premature aging, for two reasons. One is that personal genetic nutrition requires vitamin and mineral manu-facturing processes that go beyond standard manufacturing techniques.

ANSWERS TO COMMON QUESTIONS ABOUT AGING AND SUPPLEMENT USE

Let's look at three commonly asked questions and how they relate to aging and supplement use.

CAN I TAKE TOO MANY ANTIOXIDANTS?

Yes. You can take too many antioxidants if you don't know what you're doing. In this case you create a condition called "bad oxidative stress." Instead of the antioxidants neutralizing the free radicals in a step-wise fashion, they build up in the neutralization process and create a block in the step-down procedure. So the normal decrease in energy in the free radicals doesn't occur. (See accompanying figure for an illustration of the "step-like" procedure of oxidation neutralization.) This is similar to putting a barrier on a descending stairway that a lot of people are trying to walk down, blocking further descent. This oxidative stress creates more free radicals and more DNA damage, resulting in accelerated aging.

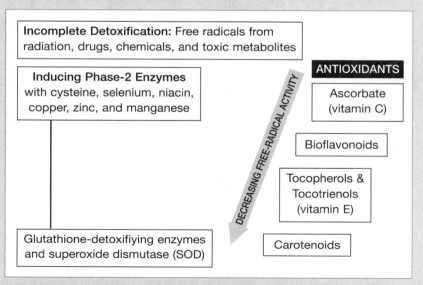

Incomplete Detoxification: Free radicals from radiation, drugs, chemicals, and toxic metabolites

Inducing Phase-2 Enzymes with cysteine, selenium, niacin, copper, zinc, and manganese

DECREASING FREE-RADICAL ACTIVITY

ANTIOXIDANTS

Ascorbate (vitamin C)

Bioflavonoids

Tocopherols & Tocotrienols (vitamin E)

Glutathione-detoxifiying enzymes and superoxide dismutase (SOD)

Carotenoids

Detox Cascade: An individual can create "bad oxidative stress" by taking too many antioxidants that create a block in the neutralization process so the normal release of free radicals doesn't occur.

Is One Supplement Tablet a Day Really Enough?

No. It is impossible to get all the vitamins, minerals, and cofactors you need to age optimally by taking one tablet a day. Your body needs a relatively constant level of key supplements throughout the day and, most important, different categories of supplements in the morning as opposed to the evening. You should take supplements for anti-aging and optimal cell health at least two times a day with the correct combinations of different vitamins and minerals.

Why Are Different Supplements Needed in the A.M. and P.M.?

The reason behind different vitamin and mineral combinations in the A.M. and P.M. has to do with the regulation of what is called the *autonomic nervous system,* or the system that is responsible for our biological clock or "bio clock." Our bio clock regulates the release of key hormones throughout the day, which helps to maintain essential bio rhythms that enhance cell repair. This bio clock also helps keep us alert in the morning as well as relaxed and ready for sleep in the evening. We have seen many patients who are taking the wrong combination of supplements and cannot get started in the morning and can't get to sleep in the evening. This loss of normal biological rhythm results in disturbances in health and aging as well as poor functioning throughout the day. It is also one of the most common complaints we see as anti-aging physicians; patients say things like "I can't get to sleep or stay asleep" or "I'm tired all day." You need to check what supplements you're taking and when. Our bet is you deregulated your normal bio clock rhythm with the wrong supplement combination.

Many companies make a fatal error when manufacturing supplements: they *press the ingredients* under high pressure to create caplets or capsules. This usually generates heat that destroys proteins and other natural compounds, rendering these otherwise good supplements useless from the cellular point of view. A pharmaceutical grade manufacturing process is required from start to finish.

The second important reason for top quality supplements needed for excellent genetic nutrition is bioavailability. If you're not able to digest the supplements rapidly—that is, if they're not made to readily absorb water so their ingredients can be released in the small intestine—they will pass through the intestine, creating nothing more than *expensive bowel movements*. Hardly the state of the art in anti-aging.

PURPOSE OF THE PGH GENE MAKEOVER CORE NUTRITION PROGRAM

Many people go to their vitamin store and buy many bottles of many different types of supplements. They then take one of each. This is known as *mega-dosing*. However, they do not know what they are taking, how much they should be taking, and whether or not negative reactions will occur in the body due to mixing ingredients that do not agree or do not work together in harmony. Negative reactions in the body between incompatible ingredients can cause more oxidative stress and DNA damage and work against your original intention to improve health.

A key component of the Personal Genetic Health Program is ingesting the proper selection and adequate amounts of key nutraceuticals in supplement form. This makes up the core nutrition supplement program that meets every individual's basic genetic needs. The PGH Gene Makeover supplement guidelines provide information about the best ingredients, the effective amounts, and gene-smart combinations. In this way the ingredients in your supplements work synergistically together to prevent unwanted negative metabolic reactions and promote positive aging reactions.

The PGH Gene Makeover core supplement is not just a multivitamin with only the essential vitamins and minerals. It can actually be classified as a *nutritional food*. It incorporates ingredients that address the following essential categories:

- Anti-inflammatories
- Botanical antioxidants
- Methylating factors
- Absorption enhancers
- Whole foods
- Blood sugar and insulin support
- DNA repair
- Fat metabolizers
- Brain enhancers
- Cellular energizers

- Amino acids
- Probiotic complex
- Vitamins
- Nucleotides–precursors for gene expression

- Fatty acid complex
- Digestive enzymes
- Minerals
- CAEs for DNA repair

The PGH Gene Makeover core nutrition program also targets DNA repair with naturally occurring CAEs (carboxy alkyl esters). The CAEs we recommend are made from a proprietary extract from the tropical rain-forest plant known as cat's claw (*Uncaria tomentosa*). But you can't go to any vitamin store, buy cat's claw, and receive the same benefit. CAEs is actually one of twenty-seven compounds found in cat's claw bark, which is extracted using a water process invented by anti-aging scientist Dr. Ronald Pero. This compound has been clinically demonstrated in two human studies to promote the body's natural ability to repair DNA. (You can locate products with this unique, patented ingredient through The Giampapa Institute, which is listed in the Resources section.)

Some people think taking vitamins just creates "expensive urine"— that is, vitamins pass through our digestive system and then our kidneys without nutrients being absorbed. This is true with some supplements. However, the compounds in the PGH Gene Makeover formula have been specifically designed to avoid such problems. The PGH Gene Makeover formula includes a lactobacillis bacteria probiotic complex, which helps you absorb more nutrients from the food you eat. Second, the formula includes a blend of the following digestive enzymes, which help compensate for an aging, inefficient digestive and hormone system:

- Amylase
- Neutral protease

- Lactase
- Lipase

- Cellulase

These measures, taken to ensure a high absorption rate, help the body's gastrointestinal track digest more effectively not only the nutraceutical but also the nutrients from foods in the regular diet. This is the primary reason why it makes sense to take your core nutraceutical supplements with a meal.

COMPONENTS OF THE PGH GENE MAKEOVER CORE SUPPLEMENT PROGRAM

Ideally, core supplements are designed to provide nutraceutical ingredients for daytime and evening needs. For example, an A.M. formula provides the fuel to power lasting daytime energy and mental activity and regulate the sympathetic nervous system. A P.M. formula helps prepare the person for a restful night's sleep that allows for body repair, rest, and recuperation and regulates the parasympathetic system. Both formulas also address the body's general nutrition needs and that most important process of DNA repair.

In addition to the PGH Gene Makeover core nutrition guidelines for A.M. and P.M., we provide the additional supplement guidelines for the most important groups of genetic differences—the 0.1 percent of individual differences we discussed earlier in this chapter—which we have observed in new research based on calorie restriction longevity studies as well as in studies linked to age-related diseases.

Nutrients That Optimize Methylation or Gene Expression

Over the past few years, it has been documented that nutraceuticals directly influence methylation or gene expression, meaning they can activate or turn on certain genes while silencing or deactivating others. This has a direct effect on how the genetic code is expressed. Fatty acids are an important class of nutraceuticals that directly affect gene expression at the membrane surface.

The typical Western diet, with its emphasis on animal, cereal, and grain products, contains an unhealthy high ratio of omega-6 to omega-3 fatty acids. Fish and dark green vegetables are good sources of omega-3s and help keep the two fats in balance. Since dietary fat regulates gene expression, intake of the wrong fatty acids leads to changes in carbohydrate and lipid metabolism.

For many years, scientists focused on the role of hormones as regulators of gene expression; the role of fatty acids didn't emerge until more recently. Hormones rely on specific receptors for uptake that are embedded in the membrane lipid bilayer. Any distortion of membrane architecture due to unavailability of the correct fatty acids affects hormone binding and alters gene expression within the cell. In this way, fatty acids can interfere with hormone regulation of a specific gene without having a generalized effect on overall hormonal control.

Fatty Acid Supplements

From an anti-aging perspective, fatty acids alter metabolism, change cellular response, and affect cellular growth and differentiation. These effects can be either beneficial or detrimental to the aging process. For example, omega-3-mediated suppression of serum triglycerides is beneficial, whereas omega-6-mediated promotion of insulin resistance is detrimental. The anti-aging benefits of fatty acids are:

- Fatty acids modulate genetic expression of key metabolic enzymes.
- Polyunsaturated fatty acids and vitamin E reduce DNA damage.

Since fatty acids make up the lipid bilayer of your cell membranes, oxidative damage to these lipids causes a dramatic change in cell function, including cell differentiation, growth, cytokine adhesion molecule release, and eicosanoid production (anti- or proinflammatory). The following nutrients are particularly effective in stabilizing cell membranes:

- Fish oil supplements reduce the need for nonsteroidal anti-inflammatory drug (NSAID) use among patients with rheumatoid arthritis, and are effective in the treatment of atherosclerosis, thrombosis and embolic events, high triglycerides, hypertension, autoimmune disease, and allergic problems.
- DHA and fish oil supplements lower triglyceride levels.
- Omega-3 fatty acid supplements reduce the risk of cardiovascular disease and sudden death in both men and women.

Daily prescription for core fatty acid supplements should include one of the following:

- Fish oils: super omega-3s—EPA, 300 milligrams (mg); DHA, 200 mg— three capsules for cardiovascular and immune benefits

- Fish or algal oils: high DHA—500 mg for brain and nerve function—one to two capsules

- Gamma-linolenic acid (GLA): 200 to 600 mg from evening primrose or black currant oils for allergies and autoimmune conditions

- Essential fatty acid combination (omega-3, -6, -9): 1,000 to 3,000 mg for general anti-aging benefits including dry skin, brittle nails, and thinning hair

B Vitamins

Nine B vitamins are considered essential for human nutrition: thiamine (B_1), riboflavin (B_2), niacin (B_3), pantothenic acid (B_5), pyridoxine (B_6), folic acid (B_9), cobalamin (B_{12}), biotin, and choline. Two others, inositol and para-aminobenzoic acid, are often included in supplement formulas. It is important to supplement all the essential B vitamins because they cofactor enzymatic reactions in groups. However, it isn't necessary to supplement them in the same amount. As noted at the beginning of this chapter, niacin, folic acid, cobalamin, and pyridoxine are the most important in protecting DNA. However, the other B vitamins are just as important because they play vital roles in gene expression and cellular metabolism.

B vitamins are water soluble and are rapidly flushed from the system, so it is necessary to supply them daily. It isn't necessary to mega-dose, but it is important to get a balance of the entire B complex, preferably above the RDI. In some instances, higher doses of individual B vitamins may be required to overcome specific deficiencies. (Optimum amounts of each B vitamin are given in the summary table of anti-aging supplements at the end of the chapter.) Now let's look at the scientific overview of the health effects of B vitamins:

- Nicotinamide (niacinamide) reverses aging mechanisms through possible modulation of histone acetylation, which is an important process that promotes gene activity.

- Niacin deficiency increases the rate of tumors in rats.

- Niacin supplementation increases DNA repair enzymes and helps overcome chemotherapy-induced damage.

- B vitamin and trace mineral supplements boost immunity in the elderly.

- B vitamins improve vascular function.

- Folate and vitamin B_6 may be chemopreventive against breast cancer, particularly alcohol-related breast cancer.

- Folic acid may protect against colon cancer.

- Folic acid, B_6, and B_{12} may lower homocysteine levels, a risk factor for cardiovascular disease.

- Folic acid may protect against Alzheimer's disease.

- Vitamin B_{12} may help resist HIV disease progression.

- Choline as alpha-glycerylphosphorylcholine (alpha-GPC) enhances overall brain function in young and older subjects.

- Alpha-GPC plays an important role in production of certain hormones and neurotransmitters such as acetylcholine and human growth hormone.

Daily prescription for core B vitamins. Take all of these supplements for optimal effect:

- Vitamins B_1, B_2, and B_6: 10 to 100 mg taken twice daily, morning and evening in divided doses

- Niacin (B_3): up to 300 mg taken as niacinamide or flush-free niacin, which does not cause the common temporary niacin response of skin reddening and itching

- Pantothenic acid (B_5): up to 1,000 mg for lipid disorders and stress

- Folic acid: 400 to 800 mcg (micrograms)

- Vitamin B_{12} and biotin: 150 to 300 mcg

- Alpha-GPC: 1,000 to 1,200 mg for cognition enhancement and growth hormone release

Minerals

Minerals function as cofactors for thousands of different metabolic enzymes. Zinc cofactors over 300 enzymatic reactions, including those involved in DNA repair. Calcium, phosphorus, magnesium, and several trace minerals form skeletal structures. Calcium also has an important role in membrane transport, and magnesium drives energy-producing reactions.

Minerals should always be taken together because there is the potential for depletion between competing minerals. For example, magnesium should always be taken with calcium in order to improve calcium utilization and guard against magnesium depletion. The best supplemental forms of minerals are those bound to amino acids or small peptides, which escort minerals across the intestinal barrier and into the blood.

A promising area of investigation is the role of peptides as chaperones for guiding minerals to their reactive sites. The discovery that peptide chaperones not only guide protein folding but also guide proteins to where they are needed earned the Nobel Prize in Medicine for Rockefeller University's Giinter Blobel, MD in 1999. Other scientists have shown that this same protein chaperone system guides minerals to the correct reactive site and that this system is common among yeast, plant, and animal cells. Consequently, minerals such as selenium, copper, iron, and zinc that are bound to yeast chaperones appear to be highly bioavailable in humans. Here is a summary of the benefits of minerals in the latest published studies:

- Calcium and vitamin D supplements reduce tooth loss.

- Calcium supplements improve lipid profiles in postmenopausal women.

- Calcium reduces the risk of colon cancer.

- Diabetes is a significant cause of low magnesium, which may lead to loss of appetite.

- Magnesium malate may provide energy reserves in fibromyalgia.

- Copper and zinc supplements may reduce bone loss in post-menopausal women.

- Zinc histidinate supplements are better absorbed than zinc sulfate.

- Chromium supplements are associated with reduced insulin resistance.

Daily core prescription for minerals. These supplements include:

- Calcium citrate malate chelate: 500 to 1,000 mg

- Magnesium amino acid chelate: 400 to 1,000 mg

- Zinc histidinate, glycinate, or yeast-bound: 10 to 25 mg

- Selenium monomethionate or yeast-bound: 200 mcg

- Chromium polynicotinate or yeast-bound: 200 mcg

- Manganese amino acid chelate: 5 to 10 mg

- Iodine from potassium iodate or kelp: 150 mcg

- Potassium amino acid complex: 99 mg

- Molybdenum amino acid chelate: 50 to 150 mcg

- Vanadium amino acid complex: 50 to 100 mcg

- Boron amino acid complex: 1 to 3 mg

Nutrients to Improve Immune Function and Block Inflammation

Both CAE extract and medicinal mushrooms work to improve immunity and prevent inflammation.

CAE Extract

Genetic damage alters the way your immune system works. Proinflammatory factors increase while those that reduce inflammation diminish. Blocking damage to DNA with antioxidants and enhancing its repair with CAE extract (from special cat's claw bark) reduces proinflammatory agents. Dr. Ronald Pero and his colleagues have found that CAEs are thus effective anti-inflammatory agents, particularly with regard to gastrointestinal disorders such as irritable bowel syndrome, Crohn's disease, and other inflammatory conditions.

NF-κB and elevated TNFα occupy pivotal roles in chronic inflammation and cancer progression. They also interfere with chemotherapeutic treatment for cancer. C-Med blocks NF-κB, in turn reducing TNFα and restoring apoptosis.

Other discoveries about CAEs include:

- A water-soluble extract containing CAEs reduces inflammation by inhibiting NF-κB.

- NF-κB inhibition is a new target for treating asthma.

- Chemotherapy may be enhanced by reducing NF-κB-blocked apoptosis.

Medicinal Mushrooms

Medicinal mushrooms contain phytochemicals that help reduce the effects of stress on your body and support immune function. They belong to a class of nutraceuticals known as *adaptogens* because they help the body restore homeostasis during stressful times. There are three important attributes of adaptogens:

1. They do not cause harm and do not place additional stress on the body.

2. They help the body adapt to various environmental and biological stressors.

3. They have nonspecific action, supporting all organ and regulatory systems in the body.

Several polysaccharides, lectins, and terpenoids have been isolated from fungi and studied for their adaptogenic and immune-potentiating effects. Among the most familiar mushrooms are shiitake (*Lentinus edodes*), reishi (*Ganoderma lucidum*), cordyceps (*Cordyceps sinensis*), and maitake (*Grifola frondosa*). Several lesser-known mushrooms are turkey tail (*Trametes versicolor*), split gill (*Schizophyllum commune*), mulberry yellow polypore (*Phellinus linteus*), and cinder conk (*Inonotus obliquus*). Cordyceps has been known in China for at least 1,000 years as the anti-aging mushroom, and reishi is considered sacred.

The only medicinal mushroom you'll find in the grocery store is shiitake, which makes an excellent addition to soups and other combination dishes, especially when your immune system is challenged. Maitake can be harvested from the wild but isn't sold in stores. All mushrooms are good for you, but culinary mushrooms are weak therapeutically when compared to medicinal mushrooms, which have a long history of use in Asia and are becoming increasingly popular in Western societies. Additionally, an active polysaccharide called beta-l,3-glucan has been isolated from yeast (*Saccharomyces* species) and is available as an immune-enhancing supplement. Here is a brief summary of the benefits of medicinal mushrooms and beta glucans:

- Mushroom polysaccharides have remarkable antitumor activity.
- Mushrooms normalize blood lipids, blood pressure, and blood sugar.
- Beta-glucan from maitake mushrooms may induce apoptosis in prostate cancer cells.
- Shiitake extracts have reduced cholesterol and have antiviral effects.
- Mushrooms are high in fiber and function as prebiotics, antioxidants, and antibiotics.

A case study involving seven men and seven women with an average age of sixty years showed remarkable results when CAE was combined with an extract containing several medicinal mushrooms. This was a particularly interesting study because members of the study group noted dramatic improvement in several inflammatory conditions, including allergies, pain with sleep loss, cardiovascular disease, and arthritis. The group was also tested for genetic damage and DNA repair capacity.

After only four weeks, DNA damage was reduced—in some people by half—and DNA repair capacity was increased by an average of 18 percent. Pain was reduced 35 percent while fatigue dropped 21 percent. There was also an improvement in allergic reactions, with 7 percent less occurrence of skin rash. Surprisingly, all but one person reported weight loss—an unexpected benefit. These were basically healthy people who had been taking a multivitamin and mineral formula, though some also took some medications. The results two women experienced were especially dramatic.

Ms. M. was a sixty-two-year-old woman who had suffered from rheumatoid arthritis for seventeen years. She had been on a variety of medications with marginal results. Her main complaints were low energy, constant pain, and an inability to sleep. After taking a combination of CAE and mixed mushroom extracts for four weeks, she reported near complete recovery and exceptional energy, and she lost fifteen pounds. She was absolutely delighted with her improvement, and her condition continued to improve at last interview, even though she was no longer taking the CAE and mushroom extracts.

Another sixty-two-year-old woman, Ms. F., had been suffering for fifteen years from an inflammatory disorder that pinched the nerve passing from her lower leg into her foot. The condition is known as tarsal tunnel syndrome and involves severe pain, burning, and tingling in the soles of the feet. She could relieve the pain with massage and elevating her feet, but as soon as she moved, the pain was excruciating and worsened as the day advanced. Special shoe inserts brought some relief, but she depended on nonsteroidal anti-inflammatory drugs (NSAIDs) and steroid injections to help diminish the pain.

Ms. F. began taking CAE and medicinal mushroom extracts for four weeks as part of the group study. She was amazed to find that not only was she free of foot pain for the first time in fifteen years, but her energy level was incredible. She found that she was virtually pain-free and only needed to take NSAIDs occasionally. She also lost five pounds.

Daily prescription for reducing inflammation and boosting immunity. These supplements include:

- CAE extract: 700 mg per day for one month, then maintain on 350 mg for medical conditions

- Medicinal mushroom extracts: 500 to 1,000 mg

Nutrients for Reducing Oxidative Stress

The substances in foods and supplements that can reduce oxidative stress are collectively called the antioxidants. These are primarily the colorful substances that occur in plants and also nutrients like vitamins E and C. The following reviews some of the key points about this important health promoting and anti-aging group of nutrients.

Carotenoids

There are hundreds of naturally occurring carotenoids, and some of these have been determined to be important to human health. They include alpha-, beta-, gamma-, and delta-carotenes, astaxanthin, beta-cryptoxanthin, lutein, lycopene, phyto-fluene, and zeaxanthin. Beta-carotene has been the most studied of the carotenes, and for many years its importance was attributed to being a precursor of vitamin A. Recently it has been discovered that alpha- and gamma-carotene are also precursors of vitamin A. Furthermore, scientists believe the most important benefits of all the carotenoids are their antioxidant capacity and cell-signaling activities.

The antioxidant properties of the carotenoids have been the object of intense interest among scientists. These fat-soluble phytonutrients with an affinity for membranes trap free radicals generated within the cell or attempting entry from the outside. A most remarkable attribute of carotenoids is their ability to regulate cell-to-cell communication; this is the basis for many of their anticancer and immune-boosting effects. Research shows carotenoids seem to express a preference for protecting a particular type of membrane. For example, astaxanthin, lutein, and zeaxanthin protect the eyes; lycopene protects the prostate; beta-cryptoxanthin protects joints; and carotenes protect DNA. Astaxanthin and lycopene are most protective against radiation from ultraviolet A (UVA) and ultraviolet B (UVB) rays. Here are some highlights from the scientific literature about the benefits of carotenoid supplements:

- Xanthophyll (yellow and orange pigmented) carotenoids and vitamins C and E can delay onset of macular degeneration and cataracts.

- Astaxanthin, canthaxantin, and beta-carotene have anticancer activity.

- High levels of carotenoids from supplements reduce oxidative damage to DNA.

- Carotenoids protect against free radicals.

- Astaxanthin and lycopene are more protective against UV radiation than beta-carotene.

- High serum levels of lycopene may play a role in the early prevention of atherosclerosis.

- Supplemental beta-cryptoxanthin and zinc reduce the risk of rheumatoid arthritis.

- Curcumin from turmeric scavenges nitric oxide, thus reducing inflammation and protecting against cancer.

Daily prescription for core carotenoid supplementation. Take all of these supplements for optimal effect:

- Mixed natural carotenoids: 5,000 to 10,000 IU (international units) of vitamin A activity from *Daniella salina*

- Lycopene: 2 to 5 mg from tomato seed extract

- Lutein and zeaxanthin: 2 to 5 mg from extract of marigold flowers

- Turmeric rhizome (*Curcuma longa*) 95 percent extract: 100 mg

- Astaxanthin, algal source: 1 to 10 mg

Note: The doses are appropriate for full-range carotenoid supplementation. Higher doses may be needed if individual carotenoids are selected for therapy.

Vitamin E

Vitamin E is one of the best-known antioxidants and is the principal protective agent found in cellular membranes. Natural vitamin E is a combination of eight related compounds, each with slightly different activity. These include four tocopherol forms: alpha-, beta-, gamma-, and delta-tocopherol, and four tocotrienols forms: alpha-, beta-, gamma-, and delta-tocotrienol.

Vitamin E neutralizes harmful free radicals attempting to damage the body's delicate cellular structures. In the process of free-radical neutralization, vitamin E itself becomes an intermediate type of free radical called alpha-tocopheroxyl. This may have a devastating effect on cells if other members of the cellular antioxidant network, including vitamin C, alpha-lipoic acid, nicotinamide adenine dinucleotide (NADH), and coenzyme Q_{10}, are not present. These other antioxidants restore alpha-tocopheroxyl to fully reactive alpha-tocopherol. Consequently, when several other antioxidants are present in the diet or supplemented, less vitamin E is required. It was established in the 1980s that levels of vitamin E drop as we age, necessitating higher intake to achieve anti-aging benefits. The scientific evidence shows that vitamin E:

- Lowers LDL cholesterol oxidation.

- Reduces the risk of stroke by protecting LDL cholesterol.

- Lowers the risk of heart disease.

- Improves insulin action in healthy older individuals and those with diabetes.

- Raises levels of glutathione, another antioxidant and detoxifying agent, in diabetic patients.

- Protects against exercise-induced oxidative damage.

- Slows the development of cataracts.

- Reduces inflammation and enhances immune response.

- Reduces cognitive decline in aging.

Tocotrienols

The tocotrienols have anti-aging therapeutic activity that is both similar and beyond that of the tocopherols. In fact, gamma-tocotrienol is the most potent cholesterol-lowering member of the entire E family. Recent studies have shown that tocotrienols:

- Lower a number of lipid-related risk factors, including total cholesterol, LDL, apolipoprotein B, and lipoprotein A (gamma-tocotrienol 200 mg per day lowered cholesterol 31 percent).

- Suppress inflammatory agents such as thromboxane B_2 and platelet factor 4.

- Reduce blood levels of lipid peroxides.

- Induce apoptosis and inhibit tumor growth in human breast cancer cells.

Daily core prescription for vitamin E and tocotrienols. These supplements include:

- Vitamin E from natural mixed tocopherols: 200 IU

- Natural mixed tocotrienols: 100 IU

Sulfur-Containing Antioxidants

In recent years scientists researching the numerous health benefits of antioxidant substances found in foods have identified a new type of antioxidants in certain plants. Cruciferous vegetables contain a family of sulfur compounds known collectively as glucosinolates. Among them are diindolylmethane (DIM), sulforaphane, calcium D-glucarate, and indole-3-carbinol (I3C). Since their discovery, scientists have produced standardized extracts of these cruciferous vegetable substances and used them in numerous trials that validate their health and anti-aging benefits. Their findings:

- I3C induces detoxification enzymes that help prevent breast cancer.

- Indole and thiosulfonate compounds isolated from cruciferous vegetables may prevent DNA adducts (compounds that insert themselves in between the strands of DNA within our cells) and colon cancer.

- Isothiocyanates help prevent cancers of the lung, mammary gland, esophagus, liver, small intestine, colon, and bladder.

- Calcium D-glucarate favorably alters hormone response and may help prevent cancers of the breast, prostate, and colon.

- DIM induces apoptosis and confers protection against DNA damage.

- DIM protects against prostate, cervical, and colon cancer. It also has antiandrogenic effects.

- Aged garlic extract inhibits TNFα and NF-κB, preventing damage to DNA and reducing inflammation.

- Aged garlic extract helps prevent liver damage from acetaminophen.

- Supplementation with aged garlic extract reduces production of F_2 isoprostanes, which are by-products of oxidative damage.

- AGE reduces cholesterol and homocysteine levels, which are significant risk factors in cardiovascular disease.

Daily core prescription for sulfur compounds. These supplements include:

- Extract of cruciferous vegetables: 500 to 1,000 mg, or for special needs, one of the individual actives as follows:

  - Calcium D-glucarate: 500 mg for colon protection

  - I3C (Indole-3-Carbinol (i3C): 250 mg for breast protection

  - DIM (Diindolylmethane): 150 mg for prostate protection

  - Garlic: 100 mg for cardiovascular and immune protection

Vitamin C and Polyphenols

Vitamin C is a complex of ascorbate and related compounds that have antioxidant and anticancer effects. Natural vitamin C is much more than ascorbic acid, and to get the full benefit of the vitamin, the entire complex should be taken. This ensures effectiveness with much lower doses and fewer side effects such as gastrointestinal irritation. It also reduces the prooxidant effect of large doses of ascorbic acid, particularly in the presence of iron.

Ascorbate is the primary water-soluble antioxidant and a key player in the antioxidant network whereby oxidized vitamin E is regenerated (see the section on vitamin E above). Ascorbate is the first line of defense against free radicals in body fluids.

Polyphenols are a group of natural plant substances that are part of the family of compounds called bioflavonoids. It is interesting to note the health benefits of these natural substances were first observed along with the discovery of vitamin C in the early 1900s. However it was not until decades later that researchers were able to clearly identify the health benefits of these precious substances. Polyphenols occur naturally in most plants, but some plants or plant parts, such as green tea and grape seed and skin, contain more of the health-promoting polyphenols. Some of the health benefits are:

- Vitamin C reduces cataract risk. A natural citrus extract of vitamin C is more bioavailable than ascorbic acid.

- Natural citrus extract of vitamin C may be a more vigorous free-radical scavenger than vitamin E in preventing LDL cholesterol oxidation.

- Vitamin C along with selenium inhibits protein glycation and formation of advanced glycation end products.

- Polyphenols protect against several chronic diseases, including cardio-vascular disease, cerebrovascular disease, lung and prostate cancers.

- Polyphenols can block prostate-specific antigen and aromatase, which are androgen-regulating proteins implicated in prostate cancer.

- Polyphenols block the cancer proliferative effects of xenoestrogens. They have antiestrogenic and anticancer effects.

- Green tea catechins (EGCG) have potent antiallergenic activity.

- Green tea catechins have thermogenic properties that aid weight loss.

- Grapeseed proanthocyanidin extract protects against DNA fragmentation and increased apoptotic cell death in oral keratinocytes of smokers.

- Pycnogenol, an extract from the bark of a French pine, is an effective anti-inflammatory agent.

- Resveratrol, found in some fruits, seeds, and particularly in the skin of red grapes may activate anti-aging enzymes.

- Cranberry juice supplements reverse cholesterol transport, decreasing total cholesterol and LDL.

Daily core prescription for vitamin C and polyphenols. These supplements include:

- Natural vitamin C from citrus: 500 to 1,000 mg

- Mixed flavonoids: 100 to 500 mg

- Grapeseed proanthocyanidins or pycnogenol: 10 to 50 mg

- Resveratrol extract: 2 to 20 mg

- Green tea extract standardized for EGCG: 50 to 100 mg

- Cranberry juice extract: 500 mg to prevent recurring urinary tract infections; 15 to 25 mg for antioxidant protection

Summary of Benefits from Antioxidant Combinations

The relationship of antioxidant levels to neurodegenerative diseases has been widely studied and published in numerous neurobiology and neuroscience publications. The overarching themes of most of these studies indicate the following:

- Alzheimer's patients have significantly lower levels of vitamins A, C, and E.

- Patients with vascular dementia have lower levels of vitamins A, C, and beta-carotene.

- Low levels of lycopene are associated with Parkinson's disease but not Alzheimer's or vascular dementia.

- No single antioxidant is universally low in the three conditions listed above. This last finding underscores the importance of supplying all of the antioxidants mentioned above to stave off neurodegenerative disorders.

Additionally, restoring blood levels of antioxidants may do the following:

- Inhibit TNFα, a promoter of inflammation and free-radical damage.

- Decrease the risk or severity of rheumatoid arthritis.

- Help maintain normal brain function as we age.

- Reduce the occurrence of Alzheimer's disease.

- Reduce the risk of ovarian cancer.

Decreasing Oxidative DNA Damage in Mitochondria

Oxidative damage to mitochondrial DNA is the root cause of aging and aging conditions. In addition to the previously mentioned antioxidants, recent findings have shed light about how a new class of antioxidants can provide extra protection to the cells' most vital energy-producing component, the mitochondria. Among the various supplement ingredients, coenzyme Q_{10} (CoQ_{10}), alpha-lipoic acid, and L-carnitine provide important protection and functional support of mitochondrial DNA.

Coenzyme Q_{10}

Coenzyme Q_{10}, also called ubiquinone, is found in high amounts within mitochondria as part of the energy-producing apparatus in these cellular power plants. It is generally understood that most aging conditions result from mitochondrial dysfunction due in part to insufficient CoQ_{10}, which works as an electron acceptor/proton donor in converting energy-rich molecules into ATP (adenosine tri-phosphate). Vital to life, CoQ_{10} is found in several forms in all living species. It has not been accorded vitamin status for humans because it can be synthesized within the body, but that requires twelve other micronutrients, including several B vitamins, vitamin C, and trace elements, for its conversion from the amino acid tyrosine. Interestingly, the same micronutrients are required for the manufacture of DNA bases from tyrosine. Consequently, micronutrient deficiencies can lead to low levels of CoQ_{10}, requiring that it be supplemented along with its vitamin and mineral cofactors.

Of medical interest, the major side effect of statin drugs such as Lovastatin is the reduced production and body stores of CoQ_{10}. Statins are used to reduce high cholesterol, a risk factor for cardiovascular disease. Paradoxically, low levels of CoQ_{10} can weaken heart muscle and lead to cardiomyopathy. Therefore, CoQ_{10} must be supplemented when statin drugs are used.

There are two critical roles for CoQ_{10}: namely, they act as antioxidant and bioenergetic molecules—and these roles are interdependent. Stress reduces energy and can make you sick. It has been well established that stress increases oxidative burden, requiring greater amounts of antioxidants to protect cells. When the antioxidants in the system are required for protection, fewer are available for energy production. At the same time, stress reduces the levels of micronutrients needed to increase internal production of CoQ_{10} to meet increased antioxidant demand. Additionally, CoQ_{10} levels drop as we age. Here are some scientific highlights on CoQ_{10}, which has been shown to:

- Inhibit lipid oxidation in both cell membranes and low-density lipoproteins.

- Protect DNA from oxidative damage.

- Protect exercising muscle from oxidative damage.

- Prevent oxidative damage to the brain.

- Stabilize membranes and reconstitute vitamin E as an antioxidant.

- Enhance cardiac function and speed recovery from heart attack; effective as an adjunct therapy during cardiac surgery.

- Maintain healthy apoptosis, and may help prevent cancer.

- Prevent thyroid disorders.

L-Carnitine (N-Acetyl-L-Carnitine)

The energetic effects of CoQ_{10} are enhanced by supplementation with L-carnitine (N-acetyl-L-carnitine). Carnitine is considered a conditionally essential nutrient for mitochondrial energetics. Carnitine transports fatty acids across membranes so that they can be converted into energy. Since heart muscle relies on fatty acids as a source of fuel, conditional carnitine deficiency reduces heart function. Carnitine is synthesized within the body from amino acids, particularly S-adenosyl methionine and lysine, with vitamin B_6, niacin, iron, and ascorbate required for its synthesis. Consequently, carnitine deficiency may occur along with other micronutrient deficiencies. N-acetyl-L-carnitine can be considered the coenzyme form of the nutrient carnitine and is preferred for maintaining brain and nerve function during aging.

Alpha-Lipoic Acid

Alpha-lipoic acid is both water and lipid soluble, making it a universal antioxidant that provides protection throughout the body. Alpha-lipoic acid is an important member of the antioxidant network, restoring both water (vitamin C) and lipid-soluble (vitamin E and CoQ_{10}) antioxidants to full scavenging capacity. Lipoic acid can regenerate itself using a niacin coenzyme (NADH). Lipoic acid is also a vital coenzyme in the conversion of glucose to cellular energy, thus helping to maintain blood glucose balance.

Lipoic acid protects DNA by sequestering free-radical-generating metals, and it has positive effects on gene expression. It is readily absorbed through the skin as well as the digestive system. Some health benefits of lipoic acid include:

- Protects both lipids and aqueous cellular components.

- Blocks NF-κB (nuclear factor-kappa beta) binding to DNA. (While NF-κB is a key compound affecting the DNA repair process, it affects the process in a negative way, so it must be blocked for DNA repair.)

- Improves carbohydrate metabolism.

- Reduces insulin resistance in muscles.

- Reduces harmful diabetic effects on red blood cell lipid membranes.

- Reverses memory loss by reducing DNA/RNA oxidation.

- Increases brain energy and skeletal muscle performance.

Daily core prescription for CoQ_{10}, carnitine, and alpha-lipoic acid. These supplements include:

- CoQ_{10}: 30 to 280 mg

- L-carnitine fumerate or acetyl-L-carnitine: 50 to 100 mg for general anti-aging benefits and as part of a comprehensive antioxidant supplement

- Alpha-lipoic acid: 500 to 1,000 mg for diabetes or diabetic complications or to reduce advanced glycation end products

Note: When these energetic nutrients are combined, less of each is required.

Nutrients for Enhancing DNA Repair

A natural substance found to enhance DNA repair is CAE (a special cat's claw bark extract) with a combination of micronutrients such as niacinamide, zinc, and carotenoids. DNA repair is critical to slowing the aging process and reducing age-related conditions. Our colleague, Dr. Ronald Pero and his coworkers in the Department of Biochemistry, University of Lund, Sweden, as well as other researchers have found that this special CAE extract:

- Enhances immune function without toxicity.

- Enhances antibody response to vaccination.

- Induces apoptosis and slows growth of leukemic cells.

- Enhances DNA repair with niacin, carotenes, and zinc.

- Nonselectively increases immune cells.

The following case studies illustrate how CAE can relieve aging conditions stemming from inflammation and lowered immune response.

Mr. R. is sixty years old and had surgery to remove herniated disks. Following the surgery he was bothered by sciatica that persisted for eight years. He tried several medications to relax his muscles and reduce inflammation, but none worked. He began taking the extract containing CAE (350 mg) twice a day. Soon after beginning the supplement, he noticed a great improvement in his flexibility and reduction in stiffness. As long as he took the supplement, he experienced no pain or sciatica. However, when he stopped, the symptoms returned. Once he began taking the supplement again, the pain and stiffness disappeared. Mr. R. also noticed improved resistance to colds and flu.

Mr. D., a newly retired postal worker, found a pea-size lesion on the right side of his forehead. Within six months, it grew to the size of a small egg. His physician advised surgery, but before it could be scheduled Mr. D. began taking 350 mg of the water-soluble extract containing CAE every day. The tumor began shrinking dramatically, so Mr. D. delayed the surgery. After four months, the tumor all but disappeared, so surgery wasn't needed. Mr. D. now takes a daily anti-aging multiple vitamin and core nutraceuticals that includes CAEs as preventive therapy.

Ms. R. is a thirty-one-year-old marketing professional. She was diagnosed with rheumatoid arthritis when she was nineteen. Her condition deteriorated after she turned twenty-nine, so she became dependent on several medications. These were extremely toxic—prednisone, the immune-suppressant drug Azulfadine, and Plaquenil, an antiarthritic drug. In addition, she received cortisone injections in her inflamed joints three or four times a year. Periodic flare-ups were so painfully intense that they registered at the top of the pain index, and her pain was particularly unbearable while traveling. Not surprisingly, Ms. R. had severe lifestyle limitations, including difficulty driving a car, working on the computer, and exercising.

Ms. R. took the CAE extract, 350 mg twice a day, for one year. During the year, her dependence on prednisone dropped from 20 mg a day to 7.5 mg a day. She also gradually reduced and finally discontinued all the other medications. Best of all, she reports her quality of life has improved dramatically. She is able to keep up with her job, has better mental function, and feels she has gotten her life back to prearthritis days.

Ms. F. is a seventy-year-old woman with rheumatoid arthritis and borderline systemic lupus erythematosus, from which she has suffered for many years. Most of the time she is lethargic, fatigued, and in constant pain due to swollen, misshapen, and inflamed joints. The immune-suppressing drugs she takes leave her with little ability to fight off frequent colds and flu, and she is developing glaucoma and macular degeneration.

Ms. F. has been taking the CAE extract (350 mg twice daily) for eighteen months. Shortly after starting the supplement, she noticed a dramatic decrease in pain and inflammation. She has not had a cold or flu since beginning supplementation. Her physician has not only noted a dramatic reduction in her symptoms but also an improvement in her eyes. The doctor attributes the latter to reduction in her rheumatoid symptoms.

Mr. H. is a thirty-year-old retailer who has had ulcerative colitis for over ten years with frequent cramping, diarrhea, and bleeding. A sigmoidoscopy exam revealed that his ulcers extended two feet into his colon. His physician originally prescribed 500 mg per day of sulfasalazine, but the dose has been repeatedly increased in order to stop more and more inflammation. With the situation becoming progressively worse, Mr. H. decided to seek an alternative treatment for his condition.

At first he took the CAE extract (200 mg per day) along with sulfa-salazine. Mr. H. noticed almost immediate relief in his condition and a reduction in inflammatory flare-ups. Not only did his diarrhea stop, but there was no longer blood in his stools. Two months after beginning CAEs, he had a colonoscopy, which showed a much smaller area of inflam-mation. Mr. H. continues to take CAEs along with a very small dose of sulfasalazine to keep his condition in check. He has more energy and a positive outlook on life.

Daily core prescription for enhancing DNA repair. These supplements include:

- CAE extract: 350 mg

- Niacinamide: 100 to 300 mg

- Zinc (amino acid chelated): 10 to 25 mg in addition to a balance of other amino acid-chelated minerals

THE PGH GENE MAKEOVER PRESCRIPTION SUMMARY

Let's summarize what a comprehensive anti-aging multivitamin and miner-al combination should contain. The suggested supplements are available in anti-aging formulas that contain many of the extras considered impor-tant for a comprehensive anti-aging program.

Top-notch formulas that contain amino acid-chelated minerals in lev-els close to the RDI will require six to eight tablets or capsules taken two to three times a day. The reason for so many pills is that minerals, partic-ularly calcium and magnesium, are required in large doses. Minerals also require great amounts of amino acids for proper chelation. In the case of calcium and magnesium, the ratio of amino acid to mineral is approxi-mately 4:1. This means that 200 mg of magnesium will require 1,000 mg of magnesium amino acid chelate. The average capsule only holds 750 mg. Specialized multiple formulas or packets with anti-aging nutraceuticals require four capsules or tablets to be taken three times a day. Table 6.2 shows the many anti-aging ingredients in the daily multiple PGH Gene Makeover supplement, which is supplied in two tablets twice daily.

Table 6.2. The Anti-Aging Daily Multiple PGH Gene Makeover Supplement Recommendations

Nutraceutical	Daily Amounts
OPTIMIZING METHYLATION OR GENE EXPRESSION	
Vitamins B_1, B_2, and B_6	10–100 mg each
Pantothenic acid	Up to 1,000 mg
Folic acid	400–800 mcg
Vitamin B_{12} and biotin	150–300 mcg each
Calcium citrate and malate chelate	500–1,000 mg
Iodine (potassium iodate or kelp)	150 mcg
Magnesium amino acid chelate	400–1,000 mg
Zinc histidinate or glycinate (listed above)	10–25 mg
Selenium monomethionate	100–200 mcg
Copper amino acid chelate	0.5–2 mg
Manganese amino acid chelate	5–10 mg
Chromium amino acid chelate or polynicotinate	100–200 mcg
Molybdenum amino acid chelate	50–150 mcg
Potassium amino acid	9 mg
Boron amino acid chelate	1–3 mg
Vanadium amino acid chelate	50–100 mcg
ANTIOXIDANTS TO DECREASE DNA DAMAGE AND OXIDATIVE STRESS	
Vitamin A	2,500–5,000 IU
Natural carotene mix (alpha-, beta-, and gamma carotenes, yielding vitamin A activity) plus lutein, zeaxanthin, lycopene, and astaxanthin	5,000–10,000 IU
Vitamin E, blend of natural tocopherols and tocotrienols	200–400 IU
Vitamin K	50–400 mg
Vitamin D	200–400 IU
Vitamin C (natural citrus extract, ascorbyl palmitate, and ascorbate)	500–1,000 mg
Botanical antioxidants (green tea, garlic, bio-flavonoids, anthocyanins, resveratrol, quercetin, and chrysin)	100–200 mg

DNA Repair and Reducing Inflammation	
CAE extract (this may require a separate supplement)	350 mg
Zinc histidinate or glycinate	10–25 mg
Niacinamide	100 mg up to 300 mg

The anti-aging multiple is for everyone. Depending on the results of your gene SNP screening test described in Chapter 7, you may need additional amounts of some supplements for your personal anti-aging program as shown in Tables 6.3 through 6.8.

Optimizing Gene Expression

Fatty acid supplements are also part of your anti-aging prescription; however, they are not included in multiple formulas such as the one described above because they are oils and require soft gel encapsulation.

Spectrum Essential Oils offers a liquid formula that provides essential fatty acids you can take by the teaspoonful or add to beverages or food.

We suggest five different formulas as noted in Table 6.3. The fish oil omega-3 blend of EPA and DHA is for cardiovascular protection. The second omega-3 blend is for brain, nerve, and eye function. The blend of omega-3, -6, -9 is for general fatty acid supplementation. Gamma-linolenic acid (GLA), an omega-6 fatty acid, is recommended for allergies, rhinitis, or eczema.

Table 6.3. Promoting Gene Expression	
Nutraceuticals	Daily Amount
Fish omega-3 blend of EPA (300 mg), DHA (200 mg)	1,000–3,000 mg
Omega-3 blend of EPA (200 mg), DHA (500 mg), or	500–1,000 mg
Omega-3 DHA alone	500 mg
Omega-3, -6, -9 blend from flax, borage seed oils	1,000–3,000 mg
Omega-6 GLA from evening primrose or borage oil	200–600 mg

Certain segments of your genes are masked at various stages of your life. However, overmethylation can silence genes that should be expressed. Consequently, methylating agents can help demethylate these vital sections of DNA and allow proper expression of the information they contain. Some of the most effective methylating agents, vitamin B_{12} and folic acid, are already included in our optimum multiple vitamin and mineral formula. Other important methylating agents are listed in Table 6.4.

Table 6.4. Methylating Agents	
Nutraceuticals	Daily Amounts
Alpha-GPC (glycerylphosphocholine) or choline	Up to 1,000 mg
S-adenosyl methionine (SAMe)	200–1,200 mg
Trimethylglycine (TMG)	100–200 mg
Dimethyaminoethanol (DMAE)	500–1,500 mg

The nutrients in Table 6.4 improve cognitive function, memory, and mental acuity and are often found in brain enhancement and mood-stabilizing formulas.

Blocking Inflammation

CAE extract is the primary nutraceutical for blocking inflammation because it inhibits NF-κB. Boswellin and curcumin are two popular Indian herbs that inhibit the pro-inflammatory COX-2 mediator. They have a long history of relieving joint inflammation and enhancing the anti-inflammatory effects of CAEs. Boswellin is available as a topical cream, and both herbs are available in capsules. (See Table 6.5.)

Table 6.5. Anti-Inflammatory Herbs	
Nutraceutical	Daily Amount
Boswellin (*Boswellia serrata*) 70% extract capsules	200 mg three times daily
Boswellin cream 5% with capsaicin or methyl salicylate	Three times daily
Curcumin (*Curcuma longa*) 95% extract capsules	250 mg three times daily

Enhancing Immune Function

Other anti-aging nutraceuticals such as probiotics (acidophilus), prebiotics (arabinogalactans or fructooligosaccharides), amino acids, and herbs are included in the formula shown in Table 6.6.

Table 6.6. Immune Enhancers	
Nutraceutical	Daily Amount
Digestive enzyme blend, pancreatin (8 x concentration) or fungal enzymes	300–500 mg
Acidophilus blend (probiotic) minimum	3 billion units
Arabinogalactans (*Larix occidentalis*) or fructooligosaccharides (FOS) (prebiotic)	200–500 mg
Medicinal mushrooms	500–1,000 mg

Additional Antioxidants

The antioxidants in Table 6.7 are essential for preserving mitochondrial function as we age. They are protective of the cardiovascular system (preventing angina, hypertension, vascular and heart disease) and must be supplemented by those taking statin medications. Alpha-lipoic acid and CoQ_{10} supplements are also indicated for prevention of protein glycation, oxidative stress, and diabetes. Natural synergism exists between L-carnitine and CoQ_{10} that increases the benefits when they are taken together.

Table 6.7. Mitochondrial Antioxidants	
Nutraceutical	Daily Amount
CoQ_{10}	60-120 mg
Alpha-lipoic acid	50–100 mg
L-carnitine or N-acetyl-carnitine	50–100 mg

Sulfur antioxidants alter gene expression by blocking enzymes that promote tumors. Their activity is enhanced by vitamin C and some flavonoids. Certain extracts from cruciferous vegetables have specific protective effects. Garlic has antiviral and antibacterial properties. (See Table 6.8.)

Table 6.8. Sulfur Antioxidants	
Nutraceutical	**Daily Amount**
Cruciferous vegetable extract	500–1,000 mg
Calcium D-glucarate	500 mg
Diindolylmethane (DIM)	150 mg
Aged garlic extract	100–500 mg

TAKING PGH GENE MAKEOVER SUPPLEMENTS

It is possible, of course, to obtain most of the supplements listed in this chapter from a number of nutritional companies. Many companies can be located by performing an Internet search for the specific supplement. If you intend to start your own PGH Gene Makeover program, however, there are some important reasons why we recommend that you use the PGH Gene Makeover supplements designed for this program. First of all, it is difficult to know the true potency of many of the products sold by the numerous companies that claim to provide quality supplements— indeed, it can be difficult to determine the reliability of the companies themselves.

Another important reason is that the PGH Gene Makeover supplements have been specifically formulated for the program. This saves both time and money for PGH Gene Makeover program participants. If you had to take each of these ingredients in separate pills or capsules, it would be difficult to swallow them all twice a day. Even if you did, eating a reasonable meal at the same time would make the effort even more difficult. It would take a long time to store and organize the many different bottles— not to mention to remove and actually take each of the tablets—more than most people would be willing to do twice a day. And we have calculated that the cost of buying the same amount of each neutraceutical separately would be close to double the cost of those that are designed for the PGH Gene Makeover core nutrition program. All these reasons make the complete PGH Gene Makeover Program the easiest, fastest, most economical way to accomplish optimal cell health.

7

The Full Personal Genetic Health Program: The Gene Makeover Completed

*Every man takes the limits of his own field of vision
for the limits of the world.*

—Arthur Schopenhauer

Let's summarize the key concepts we have covered so far in this book. To begin, various scientists, including geneticists and anti-aging physicians, have proven they can slow aging and improve the period of optimal health in mammals. Recent studies of chimps, for example, reveal they are 99.9 percent genetically identical to humans. Startling new research in both humans and chimps shows that calorie restriction changes how genes respond and creates a positive environment for aging. We also discussed how the environment, both internal and external, is the essential factor in controlling your genes and your DNA.

We now can measure the group of genes involved in human calorie restriction to determine how efficiently your personal genes are functioning. What we measure is your potential gene activity, which regulates the key cellular health processes of methylation, inflammation, glycation, oxidation, and DNA repair. Once we know what your gene efficiency is—you should know that no one's is 100 percent—we can recommend specific nutraceutical supplements that will interact with your genes and their related cellular pathways to promote more positive aging and mimic the effect of calorie restriction at the cellular level. In this way, we can nutritionally compensate for the inefficiencies you inherited with your genes.

This process is based on the new science of epigenetics, which you can take advantage of by following the PGH Gene Makeover Program.

THE PGH GENE MAKEOVER EFFECT

Following the PGH Gene Makeover Program changes the cells' internal environment so that it mimics the effects of calorie restriction without the need to actually radically restrict your caloric intake. This replicates at the cellular level the only changes that have been scientifically documented to slow aging and improve the quality of your health as well as extend your life span. Improving the five cellular processes may also slow many age-related diseases like type 2 diabetes, cancers, macular degeneration, Alzheimer's disease, cardiovascular diseases, obesity, and arthritis.

Now you know that aging does not have to be the same for you as it was for your parents. If you follow the supplement guidelines and the PGH Gene Makeover Program outlined in previous chapters, the aging process should be much more efficient and, therefore, much slower for you. Starting the PGH Gene Makeover program should be an easy decision, now that you understand the purpose of taking supplements and the importance of making lifestyle changes that improve your diet, eliminate stress, and include daily exercise. You also know how your thoughts affect how your DNA functions, and how they are transmitted through electromagnetic fields you produce with your emotions. Now you understand that anti-aging is all about doing everything possible to allow your genes to function at their highest potential in an optimal environment. Just remember:

Genes + Environment = Quality of your health

All these changes will make your Personal Genetic Health Program work faster and more efficiently.

The decided advantage the PGH Gene Makeover medical breakthrough offers you is a customized health program based on your individual genetic needs. We do this with convenient and economical home testing kits. We use the Internet to bring your test results directly to you, as well as to send you ongoing useful information about the optimal aging program and the most recent and relevant research.

THE PGH GENE MAKEOVER PROGRAM

Here's a step-by-step outline of how the PGH Gene Makeover Program works. Getting started is easy. A small brush, like a toothbrush, is used to take a small sample of skin from your inner cheek. Scraping the inside of your cheek removes some of the cells on the inner lining of the cheek that contain DNA. This sample is sent to the laboratory, where your genes are analyzed for the five essential cellular processes that are involved in age-related diseases and affected by calorie restriction. (See Figure 7.1.) For a quick review of the five processes, see "Summary of the Five Key Processes That Affect DNA Function and Are Controlled by the PGH Gene Makeover Program," on page 120.

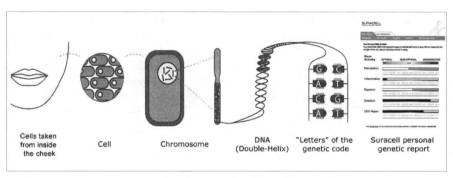

Figure 7.1. Personalized genetic health and age management
starts with a personal genetic profile.

At the same time, a small urine test is taken to measure your key bio-markers of DNA damage and free radicals. You also fill out a questionnaire that provides an indication of your environment, which influences your recommended age management supplement regimen. (See "Environmental Questionnaire.") The urine test is repeated every four months to give you objective and subjective proof that your PGH Gene Makeover Program is working. The test is repeated at this interval because every bit of the six feet of DNA in every one of the 100 trillion cells in the human body is renewed every four months, making this the ideal time period for testing.

SUMMARY OF THE FIVE KEY PROCESSES THAT AFFECT DNA FUNCTION AND ARE CONTROLLED BY THE PGH GENE MAKEOVER PROGRAM

Methylation (related to heart and vascular health): the cellular process whereby certain genes are activated and deactivated.

Inflammation (related to bone and joint health): the cellular process by which the body tries to heal itself from injury, infection, and wear and tear.

Glycation (related to blood sugar and body fat control): the cellular process in which glucose molecules bind to protein molecules and cell receptors, rendering them unable to function.

Oxidation (related to detoxification and free radical control): the cellular process which leads to the production of free radical molecules, which are the major causes of DNA and cell membrane damage.

DNA Function (related to optimal cell function): a person's current resistance to DNA damage or his or her ability to repair damage to DNA caused by hereditary, environmental, and lifestyle factors, including pollution, stress, diet, and level of exercise.

ENVIRONMENTAL QUESTIONNAIRE

1. Do you smoke?
   __ Yes   __ No

2. Do you live or work in a large city?
   __ Yes   __ No

3. Do you usually exercise *less* than twice a week for at least thirty minutes?
   __ Yes   __ No

4. Do you have undue stress at your workplace or in your home environment? (such as stress related to your job; deadlines; boss; coworkers; caring for ill children or aging parents)
   __ Yes   __ No

5. Do you currently suffer other personal stress? (such as illness or death of loved one or close friend; marital or family problems)
   __ Yes   __ No

6. Do you regularly drink more than one serving of alcoholic beverages per day? (A serving is roughly twelve fluid ounces of beer, or four fluid ounces of wine, or one fluid ounce of spirits.)
   __ Yes   __ No

7. Do you regularly drink more than three (eight ounce) cups of coffee per day?
   __ Yes   __ No

8. Do you regularly drink more than three carbonated or sweetened drinks per day? (soda, cola, sweetened tea or coffee, Gatorade, and so on)
   __ Yes   __ No

9. Have you been advised to lose weight by a medical practitioner, or do you have a metabolic disease, such as diabetes (Type 1 or Type 2)?
   __ Yes   __ No

10. Do you have any bone or joint conditions, such as arthritis, neck pain, lower back pain, spine disease, recent sports injuries, or recent injuries from accidents?
   __Yes   __ No

11. Do you have a direct family history (yourself, parents or siblings) of heart and vascular disease?
   __ Yes   __ No

12. Do you have a direct family history (yourself, parents or siblings) of cancer?
   __ Yes   __ No

13. Do you have a direct family history (yourself, parents or siblings) of Alzheimer's disease or severe memory loss?
   __ Yes   __ No

At the same time that you take these tests, you begin taking the essential core nutraceutical supplements outlined in Chapter 6. These kick off the rebuilding process on the cellular level.

Nutritional Loading

For the first one to three weeks, your body goes through a stage called "nutritional loading." This is the time when the essential vitamins, minerals, cofactors, and other nutraceutical ingredients that you have been lacking saturate both the cell and the fluid surrounding the cell. During that time, a remarkable improvement is made in the metabolic function inside each one of your body's 100 trillion cells. Waste product removal within the cells is also dramatically improved. What is really happening at the cellular level is a shift in how your genes are being utilized, as well as an improvement in the speed and efficiency of the biochemical pathways. This markedly affects how you feel. At the same time you'll also begin to practice all the nutritional guidelines, exercise, positive thinking, and other healthy lifestyle components of the PGH Gene Makeover Program.

Genetic Test Results

Approximately two to three weeks later you receive your DNA test results via the Internet. (See figure at right.) To understand the test results in the figure, you need to know that the left side of the bars indicates better or optimal results or normal gene function. The middle section of the bars indicates suboptimal results, or underactive gene function. The right side of the bars indicates deficient results or impaired gene function. These results reveal the genetic potential you inherited and show you which genes in the five groups are inefficient. These re-

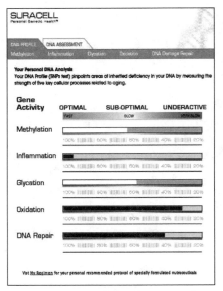

Figure 7.2. Example of a personal DNA analysis report.

sults never change so they only need to be measured once. This gene SNPs test is actually a risk assessment analysis for developing age-related diseases. It also tells us what regimens we need to start to more closely mimic calorie restriction effects at the cellular level.

Depending on your test results, a specifically designed group of pharmaceutical grade nutraceutical supplements are recommended to compensate for your less-than-efficient or suboptimal genes. Once you begin taking these individualized supplements and you continue to follow all the other components of the PGH Gene Makeover Program, the ultimate result is three-fold: (1) a decrease in DNA damage, (2) an increase in DNA repair, and (3) the beneficial effects of calorie restriction.

You also receive the results of the urine test. This shows the levels at which the five cellular processes are working and provides an objective baseline against which future tests will be measured.

Cellular Reserve

The subjective improvement you feel is obvious within the first month. The most common effect is a marked improvement in energy and sleep. Sleeping better might not seem like proof that things are improving, but adequate sleep is one of the most important components of optimal aging. It is during sleep that our body repairs most of the cell damage. It is also during sleep that key hormones are released that repair damage to DNA.

After approximately three to four months, the body enters the stage called "cellular reserve." That means the personalized nutraceutical supplements you've been taking have saturated the cell level, as well as the surrounding fluid, at an optimal state. Now even in times of stress or extra need, core nutrients and ingredients are available within the cell and its surrounding environment to fight off bacteria, viruses, pollution, toxins, or any of the many other stresses in the twenty-first-century environment.

PGH Gene Makeover Program users experience a remarkable change after the first four months. Common improvements include:

- Sustained energy throughout the day.

- More restorative sleep.

- Less mental fatigue.

- Marked improvement in skin quality (more hydration, fewer wrinkles, less slackness).

- Improved body composition (less body fat).

- More positive mood and fewer mental/emotional mood swings.

- Improved resistance to colds and flu.

Measuring Genetic Protection Factors: Your GPF

After four months, you take another urine test at home to document that the program is making a positive difference in how you age and how you feel. We call these measurements "genetic protection factors" (GPF), which we introduced in Chapter 3. Changes in DNA damage and free-radical levels indicate a number of things. The lower the DNA damage rates, the better your cellular copies have become. Because free radicals are the main cause of DNA damage at the cell membrane level, the lower the free-radical rate is, the better your body is functioning. Less damage is occurring at the cellular level.

Testing for the two key biomarkers is repeated with a urine test approximately every four months to determine your GPF and the program's progress. The biomarkers provide essential objective evidence that you are improving at the DNA level. (See Figure 7.3.)

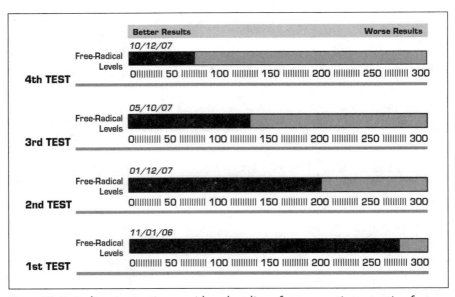

Figure 7.3. Periodic urine testing provides a baseline of your genetic protection factors. The tests measure DNA damage and free-radical levels at the start of the program and monitor ongoing progress.

At the same time, you fill out another questionnaire to determine how you feel on the subjective level. (See "Subjective Age Management Questionnaire.") This self-evaluating questionnaire allows you to determine your improvement over time for such things as energy, sleep, and the quality of your day.

SUBJECTIVE AGE MANAGEMENT QUESTIONNAIRE

Name:_____ Date:_____

Months on Program:_____

Please rate each item in Parts 1 through 3 by giving it a number in the Negative column (if you are worse than at the beginning of the program) or the Positive column (if you improved from the beginning of the program). If no change is observed, mark a zero. To rate each item, evaluate it in 20 percent increments from 1 to 5 with 1 being a barely noticeable change (20 percent) and 5 being an extremely noticeable change (100 percent).

PART 1	Negative Change	0	Positive Change
Mental Functions			
Daily energy levels			
Sense of alertness			
Memory			
Mental focus			
Overall stress level			
Skin Quality & Appearance			
Age spots on hands and face			
Skin thickness/firmness			
Body skin moisture level			
Face skin moisture level			
Nail strength			
Nail growth			
Hair strength			
Hair growth			

	Negative Change	0	Positive Change
Immune system			
Allergic reactions			
Healing			
Colds & flus			

PART 2

Before program	Time of questionnaire
Weight:	Weight:
Chest:	Chest:
Waist:	Waist:

	Negative Change	0	Positive Change
Digestive system			
Digestive health			
Bowel movement regularity			
Skeletal system			
Bone & joint pain			
Flexibility			
Joint swelling & redness			
Body composition			
Weight improvement			
Body contour			
Exercise endurance			
Exercise strength			

PART 3	**Negative Change**	**0**	**Positive Change**
Sexual status			
Libido level			
Firmer erection (men)			
Frequency of sexual relations			
Increased sensitivity			
Menopausal symptoms (women)			
Sleep function			
Ability to fall asleep			
Sleep interruption			
Sleep duration			
Dreaming sensation			
Thirst			
Water consumption			

If both test values are improving, you know for sure that there's major improvement in how your genes are working. The PGH Gene Makeover Program is reaching the deepest level of your cells where it's preventing the key symptoms of the aging process. Your genes are now working at their full potential to promote positive aging at the cellular level, which has been one of the key goals since the beginning of the program. This is the closest we can scientifically come to achieve anti-aging or age management at the present time.

Commonly Asked Questions

A number of questions are often raised when people consider whether they should adopt the PGH Gene Makeover Program.

Is It Too Late for Me to Start This Program at Seventy?

No one is too old to start this program! In fact, clinical results have shown older people see and feel the differences even more rapidly and with greater overall effect because cells have been nutritionally starved longer. Typically, over just a four-month period, cellular function can be tremendously improved and quality of life restored.

We've heard numerous times from patients in middle age that they have used the core nutrition approach with their older and confused parents in nursing homes. They've found that after three to four weeks of using the PGH Gene Makeover core nutrition program, their elderly parents previously diagnosed with a form of dementia called organic brain syndrome or early Alzheimer's disease have, after their mental fog and confusion cleared, asked why they were in a nursing home. This has even allowed a small number of people to return home where they are much more comfortable with their families.

What Is the Ideal Number of Supplements I Will Take?

The gene test will recommend on average two to three additional tablets to be taken in the A.M. and P.M. Additional tablets, for a total of anywhere from four to six throughout the day, may be required. The regimen usually looks like the following:

Taken in the Morning
(ES = extra strength; RS = regular strength)
1 tablet: Oxidation/Detoxification (ES)
2 tablets: Essential Genetic Formula A.M.
1 tablet: Methylation/Heart and Vascular (RS)
1 tablet: Inflammation/Bone and Joint (RS)
1 tablet: Glycation/Blood Sugar (RS)
1 tablet: Optimal Cell Function (RS)
Total taken in the morning = 7 tablets

Taken at Bedtime

1 tablet: Oxidation/Detoxification (ES)

2 tablets: Essential Genetic Formula P.M.

Total taken at bedtime = 3 tablets

What Should I See and Feel Within the First Month of Following the PGH Gene Makeover Program?

Look to the skin first! Because the skin is the largest organ in the body, some of the most noticeable effects of Personal Genetic Health are initially seen here. In general, the skin starts looking less red and inflamed. The texture is smoother, pore size is much less noticeable, and the skin appears more hydrated or tight.

Although changes are subtle, most PGH Gene Makeover Program users get comments from friends and relatives like "Hey, you look rested" or "You look younger." They can see something has changed, but they can't put a finger on it. The changes are mainly due to a decrease in inflammation and free radicals that damage fibroblasts responsible for manufacturing collagen and hyaluronic acid—the gel substance between the stem cells that holds water hydrating the skin. The decrease in free-radical levels causes a drop in DNA damage and more rapid skin turnover.

While we're talking about changes, let's correct some misperceptions about skin perpetrated by the billion-dollar cosmetics industry. It is a biological fact that 98 percent of skin nourishment comes from the inside out—from the foods, water, and supplements you ingest. The vast array of tiny blood vessels in the skin's dermal level supplies virtually all the nutrition necessary for optimal skin health. The key is getting the right nourishment into our bodies from the foods we eat and from the supplements we take—not from what we put on our skin!

The dead layer of skin on the surface of the epidermis virtually stops the vast majority of all topical skin creams from penetrating the skin's deeper layers. With the advent of new synthetic cells (microsomes and microliposomes), key ingredients are able to penetrate into the upper level of the dermis, but this still does not negate or replace the importance of optimal nutrition from food and supplements. (See Figure 7.4 on the following page.)

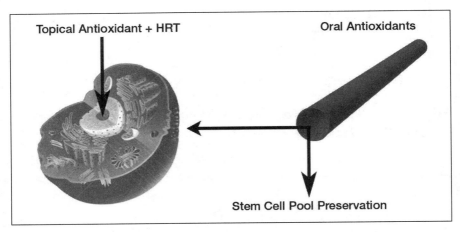

Figure 7.4. Comprehensive Skin Management Therapy
This requires a combination of the latest topical skin cream and oral antioxidants.
Stem cell pool preservation, aided by limiting DNA damage in general, is the key focus
to optimal skin health and regeneration. Over the past five years at the Giampapa
Institute the clinical use of both topical DNA repair compounds and oral core nutrition
supplements has resulted in remarkable improvement in the condition of aging skin.

A GLIMPSE OF THE POSSIBLE FUTURE AS YOU ADOPT THE PGH GENE MAKEOVER PROGRAM FOR A LONGER, HEALTHIER LIFE

Until "gene transplants" are available, the Personal Genetic Health Program is the most advanced, scientifically based program for optimal health and aging. Sometime in the not too distant future, we will be able to manipulate our adult stem cells, or perhaps reprogram our DNA. After taking the gene test discussed in this book and identifying the defects in each one of the five genetic cell processes, we will be able to introduce into a small sample of adult stem cells the genes necessary to correct inherited defects in each of the five key cellular processes, and actually recreate the gene activity we had when we were in our twenties and thirties.

Once these new genes have been edited into the normal gene combination in our adult stem cells, these adult stem cells will be multiplied and returned to our body through a simple intravenous transfusion much like a blood transfusion. No rejection reaction will occur, because these are our own cells. Within a small period of time, the new adult stem cells will create new somatic cells (discussed earlier in this book). Our genes will then function at an optimal level, mimicking the effects of calorie restriction and working efficiently for years.

The goals for the PGH Gene Makeover Program are far-reaching:

- Improve environmental effects: We can control things like diet, exercise, and mind state while minimizing exposure to toxic elements, pollution, and local radiation.

- Improve the function of the aging equation: We can control methylation, inflammation, glycation, and oxidation at the cellular level.

- Improve DNA replication and gene expression: We can improve the ratio of DNA repair over DNA damage, decreasing cell mutation and promoting more accurate cell copies during cell replication. This preserves adult stem cell pools.

- Utilize personally directed genetic nutrients: We can take specific nutraceuticals to optimize genetic expression and health span.

Even with this remarkable future technology, for which basic principles are already well established, we will still need optimal nutrition and correct ratios of carbohydrates, proteins, and fats in our meals to maintain youthful body composition and health. But until this science arrives for the vast majority of the public, the Personal Genetic Health Program will be the twenty-first-century anti-aging breakthrough—the best option we have right now for optimal health and aging.

The information in this book is radically new, but it is based on sound revolutionary scientific principles. It is our hope that what you have read and what you incorporate into your life—with a small amount of self-discipline and effort—will allow you to become "a master of your health, not a victim of your genes."

Remember:

- Information changes habits.
- Habits change actions.
- Actions change environment.
- Environment changes your genes.

You alone are responsible for you! Good luck with your Personal Genetic Health Program!

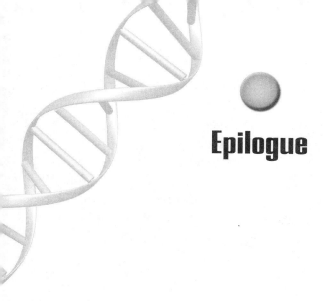

Epilogue

You have just read about the new renaissance in medicine: about helping the genes we have inherited work to their optimal potential. We now know that if we can recreate the environment in ourselves that closely emulates what happens when we eat less (restrict our calories), we create a condition in our genes that allows them to work more efficiently for ultimate Personal Genetic Health. The information in this book has emphasized how we can accomplish this condition and decrease DNA damage, by using the latest gene SNP technology and nutraceutical supplements based on our individual needs.

The next advancement in anti-aging technology and age management therapy will involve actually reprogramming our DNA with naturally occurring compounds found within our bodies as well as in nature. We call these amazing compounds Genetic Expression Modifiers, or GEMs, already being utilized at the Giampapa Institute in more advanced anti-aging programs. These compounds are capable of actually reversing the master clock of aging by silencing aging genes and re-awakening our youthful genes. GEMs also dramatically help protect us from cancer and a number of age-related diseases.

GEMs not only effect our body cells (somatic cells) but also the rest of our adult stem cells that are responsible for replacing body cells as they are damaged or normally wear out over time. Not only do GEMs help our body cells work more efficiently, resulting in less drain on our adult stem cell reservoir, but they allow them to replace our aging and damaged cells more efficiently.

Together with the revolutionary information in this book about Personal Genetic Health, GEMs are the next step in medical anti-aging technology. They will generate the most effective personalized age management program ever available. This scientific achievement will usher in a new era of health span and longevity. It will set the standard of age management therapies for decades to come. You will be able to read about this leading-edge information in my next book, as the saga of Personal Genetic Health science continues.

—Vincent Giampapa, M.D., F.A.C.S.

Gene-Smart PGH Gene Makeover Nutrition

Along with your daily dose of healthy nutraceutical supplements as discussed in Chapter 6, eating the right foods in the right amounts is vital to maximize the PGH Gene Makeover positive aging effect. It's become well known in recent years that most diets prove deficient in several essential nutrients and also are way out of balance with regard to the type and amount of essential macronutrients they contain, including protein (amino acids), fats, carbohydrates, and fiber. So eating the right foods in the right amounts will ensure adequate essential nutrient intake and nutrient balance that will affect your genes, hormone balance, health, and longevity.

Optimal nutrition differs among individuals. Due to genetic differences, individuals respond differently to nutrients and supplements. However, while there are differences between us, there are also many similarities. The nutrition guidelines in this appendix focuses on this nutritional common ground.

THE IMPORTANCE OF BODY COMPOSITION

Excess fat is the culprit for accelerated aging because fat cells cause hormonal imbalances that are bad for your health. The hormones produced by fat cells cause inflammation, excessive clogging of arteries, deterioration of the heart and vascular system, and accelerated aging. It is for this and other reasons that you must strive to maintain good body composition.

Doctors measure body composition by estimating the ratio of your body fat to lean muscle mass. It is a way to keep track of how much body fat your body contains versus other tissues, like muscle, bone and organs, which constitute your lean body mass. You may have encountered the term Body Mass Index, or BMI, as a measure of an overweight condition or obesity. While the BMI is in wide use, it has limitations in determining body composition because it does not distinguish between lean and fat components of body weight. If a person has a BMI index number higher than twenty-five, he or she is considered overweight and a person with a BMI higher than thirty is considered obese.

A more reliable, quick method to estimate body fat content is called waist circumference (WC) and waist-hip ratio (WHR). The waist measurement for both genders is taken just above the hip bone and over the navel. The calculation of WHR is waist size divided by the girth at the broadest part of the buttocks. According to the World Health Organization, men who have a WC greater than 101 cm (forty inches) and a WHR greater than 0.95 are obese, and women who have a WC greater than 88 cm (thirty-five inches) and WHR greater than 0.88 are obese. Table A.1 shows body fat ranges, waist circumferences, and waist-hip ratios used to classify people as normal, overweight, and obese.

Table A.1. Body Fat Based on Waist Circumference to Waist-Hip Ratios			
	Normal	Overweight	Obese
Female	18–22%	22–30%	>30%
Male	15–17%	17–25%	>25%
	Normal	Overweight	Obese
Female	<80 cm	80–88 cm	>88 cm
Male	<94 cm	94–101 cm	>101 cm

It is important to note that these are general ranges that should be used as guidelines; your ideal body composition should be verified by your health professional.

PGH GENE MAKEOVER NUTRITION GUIDELINES: EATING FOR ANTI-AGING BENEFITS

Anti-aging eating is very different from a weight-loss plan. While the latter focuses on losing pounds, anti-aging eating is a lifestyle modification that alters your body composition so you obtain a more youthful ratio of fat to lean muscle. By focusing on body composition instead of weight loss, you will:

• Burn calories more efficiently by increasing muscle mass.

• Decrease age-related conditions as muscle mass increases.

• Improve your quality of life as your body functions better.

• Normalize hormone levels and improve the balance between them, thus directly affecting the rate of aging.

• Increase antioxidant levels.

• Reduce pain and inflammation.

• Improve cellular function.

Figures A.1 through A.3 on the following pages show the positive and negative effects of food on various functions as the body ages:

The PGH Gene Makeover diet should have:

• Enough calories, but not too many; remember the benefits of calorie restriction. In general most people only need a 2,000- to 3,000-calorie per day diet, though athletes might need more. As we get older, especially after age fifty, we need to eat less because our body uses fewer calories.

• Balanced macronutrients. Our diet should consist of a daily intake of about 40 percent carbohydrates, 35 percent protein, and 25 percent fat.

  ○ **40 percent carbohydrate:** the best type of carbohydrate-containing, nutrient-dense, whole foods are fruits and vegetables, including grains, nuts, herbs, and spices, which do not cause a rapid rise in blood sugar levels. They have low glycemic index/glycemic load ratings.

FIGURE A.1 **Effects of Food on Hormones**

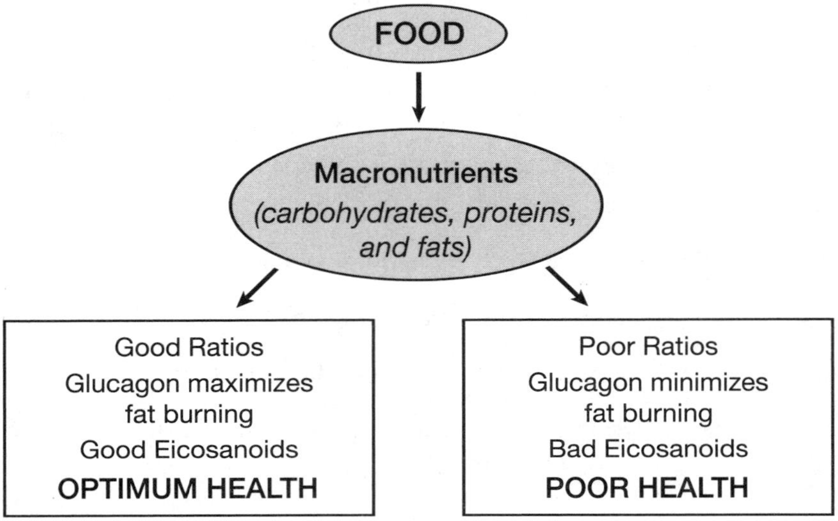

FOOD

↓

Macronutrients
(carbohydrates, proteins, and fats)

Good Ratios
Glucagon maximizes fat burning
Good Eicosanoids
OPTIMUM HEALTH

Poor Ratios
Glucagon minimizes fat burning
Bad Eicosanoids
POOR HEALTH

FIGURE A.2 **Insulin Resistance and Effects on Aging**

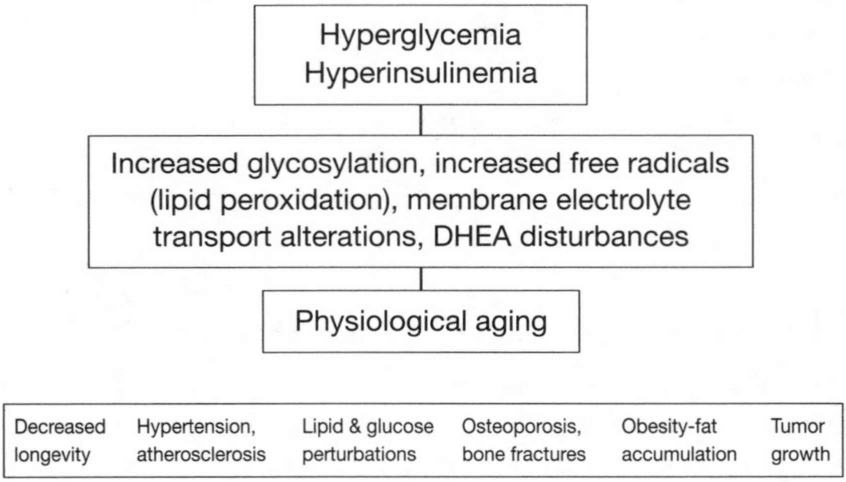

Hyperglycemia
Hyperinsulinemia

Increased glycosylation, increased free radicals (lipid peroxidation), membrane electrolyte transport alterations, DHEA disturbances

Physiological aging

| Decreased longevity | Hypertension, atherosclerosis | Lipid & glucose perturbations | Osteoporosis, bone fractures | Obesity-fat accumulation | Tumor growth |

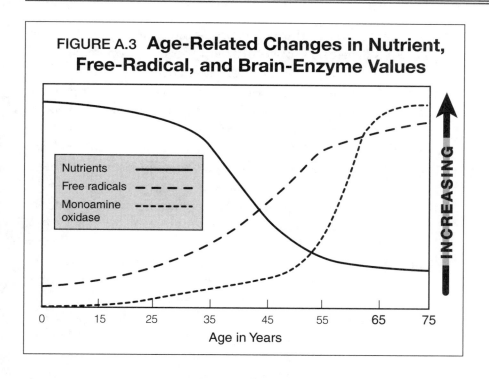

FIGURE A.3 Age-Related Changes in Nutrient, Free-Radical, and Brain-Enzyme Values

A diet rich in fruits and vegetables is rich in phytonutrients that have disease-preventing power. Phytonutrients have color-coded pigments related to specific disease-fighting antioxidant properties. For example:

— Yellow, orange, bright red: carotenoids

— Green: sulfur compounds, isothiocyanates, indols

— Purple, blue, black, magenta: phenolics, flavonoids

— Tan: phytosterols, phytoestrogens, fiber, saponins

— White: protein, omega-3 fatty acids, omega-6 fatty acids

You need to eat at least two to three foods from each of these color categories every day to promote such benefits as an increase in good cholesterol (HDL) and a decrease in bad cholesterol (LDL) and triglycerides; less chance of developing type 2 diabetes, high blood pressure, and heart disease; and increased weight loss, because lower insulin levels reduce the amount of stored fat.

○ **35 percent protein:** although the U.S. Department of Agriculture recommends 15 percent protein, that's just too low for optimum health.
Eating 35 percent protein provides better appetite control, improves
blood lipids, increases good HDL, decreases bad LDL and high levels
of triglycerides, protects bone mass (needed to prevent osteoporosis), reduces insulin resistance, increases the rate of weight loss, and
increases heart healthy vitamins such as vitamin B_{12}, folic acid, and
B_6. White proteins such as fish and chicken offer many health benefits; avoid or limit red proteins found in most meats.

It is important to increase protein intake as you age in order to
maintain muscle mass, reduce insulin resistance, and repair DNA.
However, excess intake of animal proteins makes the body more
acidic, changing the cellular environment and making it less resistant
to disease. If you increase animal protein, be sure to eat more fruits
and vegetables that are alkaline to maintain optimal acid/base (pH)
balance in the body. Remember: legumes and vegetables are excellent sources of protein that do not alter pH balance. Adding two
drops of a solution of ionized water, potassium, and sodium minerals
in hydroxide to eight- to ten-ounces of drinking water also helps stabilize pH balance. This preprepared solution can be found in many
health food stores labeled "pH stabilizers" or "alkaline boosters."

○ **25 percent fat:** good fats are vital to the PGH Gene Makeover diet,
but they must be eaten in just the right amounts. Your diet should
be high in MUFAs (monounsaturated fatty acids) like olive oil and
canola oil and PUFAs (polyunsaturated fatty acids) like omega-3 and
omega-6 fish oil. Many fatty acids can be produced in our bodies,
but essential fatty acids omega-3 and omega-6 must be provided in
the diet. They promote cellular membrane function, production of
eicosonoids, blood pressure control, and better function of our
blood platelets. The saturated fats in red meat should be avoided.

The Mediterranean-style diet that's high in monounsaturated fatty
acids from olive oil and high in whole foods, fruits, and vegetables is
increasingly popular. It's been found to protect against age-related
cognitive decline and other diseases. The best olive oil is unrefined,
dark green, extra-virgin oil whose color comes from naturally occur-

ring antioxidants. Avoid lighter oils whose antioxidants were reduced by heating or bleaching during processing.

Nuts are also sources of healthy fats when eaten in moderation. Flax seed oil and macadamia nut oil are good alternatives to olive oil. One Harvard Health Study showed that male professionals who ate nuts at least twice a week lowered their risk of heart attack by 47 percent and reduced their risk of coronary heart disease by 30 percent.

- Vitamin and mineral supplements based on your genetic makeup (as discussed in Chapter 6).

- **High fiber.** Diets high in fiber from fruits, vegetables, and whole grains have been found to improve insulin sensitivity and lower cholesterol levels.

- A regimented meal plan with healthy snacks between meals. Eating regular meals, starting with breakfast, has many beneficial effects, and snacks between meals help prevent hypoglycemia. For optimal health, don't skip breakfast, eat at regular intervals, and use nutritious meal replacement shakes and nutrition bars to guarantee high-quality nutrition intake between meals or for meals on the go.

To summarize, according to the PGH Gene Makeover food pyramid shown in Figure A.4 on page 142, you should eat eight to fourteen ounces of protein, two to three ounces of dairy, and one to two ounces of oils, with plenty of whole fruits and vegetables.

GENERAL GUIDELINES FOR PGH GENE MAKEOVER MEAL PLANNING

Keep these guidelines in mind when planning your meals:

- Eat your meals at planned, regimented times of the day in a stress-free environment

- Eat smaller, nutritious meals more frequently.

- It's unhealthy to eat on the run.

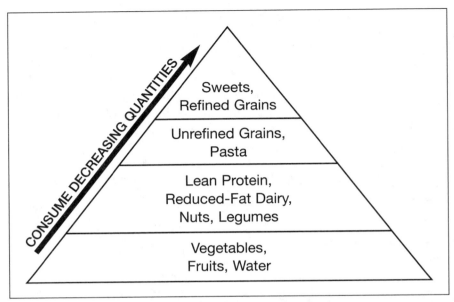

Figure A.4. The PGH Gene Makeover food pyramid.

- Emphasize protein, concentrating on vegetable sources as much as possible, including tofu and soybean products.

- Choose seasonal fruits and vegetables with the most intense color.

- Choose brown or tan carbohydrates, not white ones.

- Eat your protein foods first, followed by nonstarchy vegetables, then carbohydrates.

- Drink plenty of liquids between meals instead of with them.

- If you drink alcoholic beverages, do so with meals and in moderation.

- Use only high-quality extra-virgin olive oils.

PGH Gene Makeover Anti-Aging Exercise

*Exercise is the cheapest form of hormonal therapy available.
It's readily accessible to all at anytime and any place.
When you change your hormonal levels, you change your
gene function. That makes exercise a form of gene therapy.*

DR. VINCENT GIAMPAPA

Now that you know the health-promoting effects of the Personal Genet-ic Health nutrition program, it's time to take a look at the additional ben-efits exercise has to offer. Keeping in mind that time is precious, my colleagues and I developed the Giampapa Institute™ Personal Exercise Pro-gram that is both time sparing and convenient, so anyone can do it at home. We also wanted it to be easy and fun for adults of all ages.

JOIN THE ANTI-AGING EXERCISE REVOLUTION

Everybody can experience the life-enhancing benefits of the right exercise program. Here's why:

- Exercise, like food, is an essential way of controlling vital hormonal lev-els. It is also one way of controlling your body composition and creat-ing positive changes in your quality of life. From our research, we observed that a number of the key health biomarkers of aging are improved with exercise. For instance, exercise:

1. Lowers excess blood glucose.

2. Lowers excess insulin.

3. Elevates growth hormone.

4. Elevates testosterone.

- Exercise speeds up the fat-burning process by burning calories more efficiently and improving the metabolic rate. That makes us better at using the calories from the foods we eat and less inclined to store the food as fat. It also keeps our cells healthy, including the important energy-producing mitochondria. When the mitochondria in your cells are good at producing energy, your entire body is healthier.

- Exercise lowers insulin levels and improves the functionality of insulin. When the insulin level is lowered, your fat-burning hormones, like glucagon, have a chance to burn stored body fat for energy.

- Fat loss is promoted simply through a regular—at least four days a week—long-term walking program without any severe dieting.

- Vigorous regular walking leads to reduced body fat storage, lower insulin requirements (a 36 percent decrease in the ratio of insulin to glucose concentration), and spontaneous reduced food intake.

- Intense exercise can maintain your muscle mass, build additional muscle mass, increase your strength, and increase bone health.

TYPES OF EXERCISE AND THEIR EFFECT ON HORMONES

When you look at all the exercise programs in books and videos and at the gym, it can be bewildering. But exercise falls into two major categories: (1) aerobic, including such cardiovascular activities as jogging, walking, running, biking, stair-stepping, and using a treadmill and (2) anaerobic, including such resistive or weight-bearing activities as weightlifting and calisthenics. To summarize:

- Aerobic exercise results in insulin reduction and improved cardiovascular conditioning.

- Anaerobic exercise results in elevated secretion of human growth hormone (HGH) and testosterone to maintain or increase muscle mass, bone mass, and therefore functionality.

- Both aerobic and anaerobic exercise benefit your heart and your circulatory system.

The important age-related hormones—insulin, glucagon, human growth hormone, insulin-like growth factor-1 (IGF-1), and testosterone—are affected by both types of exercise (See Table B.1.).

Table B.1. Effects of Exercise on the Five Health Control Gene Groups and Related Hormones	
Five Age-Related Gene Groups	**Effects of Exercise**
Methylation: Turns on and off specific genes, generally related to the heart and vascular health.	Improves metabolic processes. Good HDL cholesterol is elevated; bad LDL cholesterol is decreased.
Inflammation: Creates inflammatory molecules in the cell.	Improves joint mobility, decreases inflammation, improves blood flow.
Glycation: Exerts blood sugar effects on the proteins and body fat levels.	Reduces glucose levels, improves insulin function, and decreases body fat.
Oxidation: Produces free radicals.	Increases natural antioxidant protection, such as SOD (superoxide dismutase) and catalase. Improved blood flow means better delivery of dietary antioxidants to cells and removal of waste products.
DNA Repair: Repairs damaged DNA.	Exercise reduces stress and stress hormones, like cortisol and insulin, which reduces DNA damage and telomere shortening.

It's essential to examine the effects of each type of exercise on these key hormones to appreciate the PGH Gene Makeover exercise program:

- **Aerobic (cardiovascular) exercise:** Growth hormone levels are at a low level before a period of exercise. Within thirty minutes after aerobic exercise, your pituitary gland releases a surge of growth hormone, which reaches a peak in the fifteen to twenty minutes that follow. This exercise-induced surge of growth hormone helps repair damaged muscle mass and increases or maintains muscle mass while decreasing stored body fat levels. As growth hormone levels increase with the initiation of exercise, insulin levels begin to drop and glucagon levels begin to rise. This drop in insulin levels and rise in glucagon and glucose levels help to further stimulate the release of growth hormone and testosterone. These hormonal responses occur mainly during aerobic exercise: walking, jogging, running, and biking, That's why it's more beneficial to begin aerobic exercises *before* resistive exercise or weightlifting. Aerobic exercise results in improved cardiac conditioning and pulmonary function, which are some of the most important and earliest reversible biomarkers that decrease with aging. (See Figure B.1.)

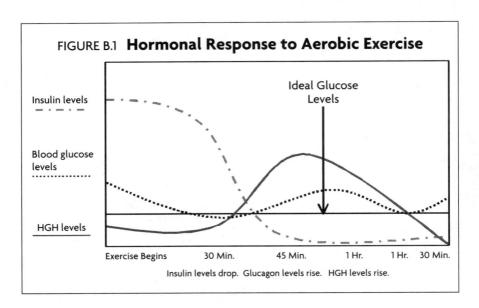

FIGURE B.1 **Hormonal Response to Aerobic Exercise**

- **Anaerobic (resistive) exercise:** Within minutes of a weightlifting session, both testosterone and growth hormone are released; after thirty minutes, they hit their peak. The effects of growth hormone and testosterone continue for approximately another forty-five to sixty minutes. It is during this period that weightlifting, which improves muscle tone, muscle mass, tendon and ligament strength, has its effect (see Figure B.2).

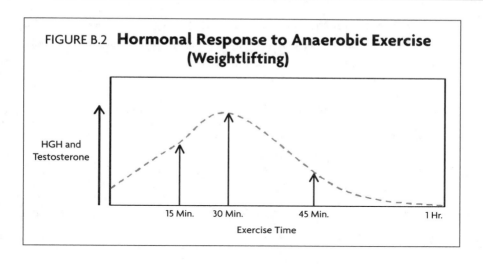

FIGURE B.2 **Hormonal Response to Anaerobic Exercise (Weightlifting)**

THE IDEAL ANTI-AGING EXERCISE PROGRAM

From Figure B.3 we can see that the ideal sequence of exercise for anti-aging purposes includes a period of approximately thirty minutes of aerobic exercise before anaerobic exercise.

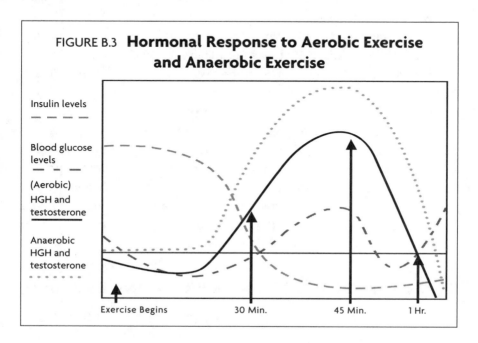

FIGURE B.3 **Hormonal Response to Aerobic Exercise and Anaerobic Exercise**

It is also important to do the right amount of the right exercise. Too much of both kinds of exercise (more than forty-five minutes) can create negative effects as far as anti-aging parameters are concerned.

- After forty-five minutes, excess levels of cortisol increase to such an extent that other hormone functions become impaired.

- Eventually even testosterone levels start dropping as testosterone precursors, like DHEA, shift to make more cortisol.

- Free-radical production also accelerates.

- Elevated cortisol causes an increase in blood glucose, which stimulates insulin production.

For ultimate health benefits, keep your exercise to a moderate duration. Do aerobic exercise at least five times a week for a minimum of twenty minutes. For ideal heart and pulmonary biomarker improvement, aerobic exercise should be performed at 55 percent to 65 percent of your maximum heart rate. An easy way to calculate that is to simply deduct your age from 220. For example, if you are forty-five, your maximum heart rate will be 220 minus 45, which equals 175. Then, to calculate your exercising heart rate, multiply 175 times 55 percent and 65 percent. For example, 175 times 0.55 equals 96 beats per minute, and 175 times 0.65 equals 114 beats per minute. So you should keep your heart rate between 96 and 114 beats per minute. Make sure to have your doctor verify that this general rule of exercising heart rate is right for you.

Do anaerobic exercise three to four times a week for no more than forty minutes. Something very interesting happens to your body's appearance from anaerobic exercise that does not occur from aerobic exercise— it makes your skin look younger. Just take a look at the bodies of those muscle men and women in magazines and on television. From the neck down the bodies of people in their sixties and seventies look as young as those of people in their twenties and thirties.

DON'T FORGET STRETCHING & BREATHING EXERCISES

Two other types of exercise are part of an age-management program. (See

Figure B.4.) Stretching is important in nervous system regulation and balance; it helps to maintain flexibility of the spine and joints, and it improves the lymphatic system and blood flow to your vital organs. It also helps spinal column alignment, which is essential to avoid nerve and disc injuries as well as lower back pain, which occur all too often with aging. Stretching should be practiced every day.

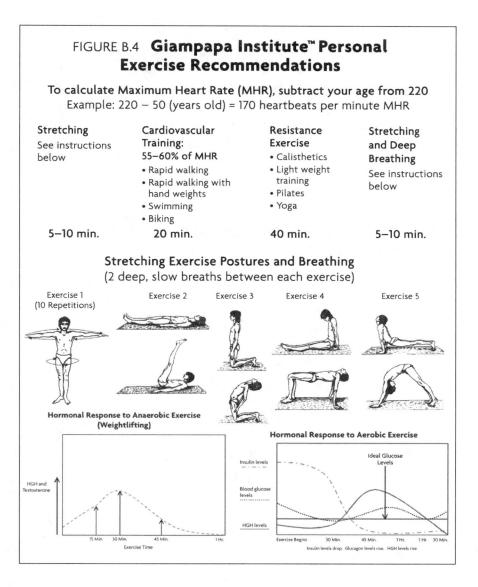

FIGURE B.4 **Giampapa Institute™ Personal Exercise Recommendations**

To calculate Maximum Heart Rate (MHR), subtract your age from 220
Example: 220 − 50 (years old) = 170 heartbeats per minute MHR

Stretching	Cardiovascular Training:	Resistance Exercise	Stretching and Deep Breathing
See instructions below	55–60% of MHR	• Calisthetics	See instructions below
	• Rapid walking	• Light weight training	
	• Rapid walking with hand weights	• Pilates	
	• Swimming	• Yoga	
	• Biking		
5–10 min.	**20 min.**	**40 min.**	**5–10 min.**

Stretching Exercise Postures and Breathing
(2 deep, slow breaths between each exercise)

Exercise 1 (10 Repetitions) Exercise 2 Exercise 3 Exercise 4 Exercise 5

Hormonal Response to Anaerobic Exercise (Weightlifting)

HGH and Testosterone

15 Min. 30 Min. 45 Min. 1 Hr.
Exercise Time

Hormonal Response to Aerobic Exercise

Insulin levels
Ideal Glucose Levels
Blood glucose levels
HGH levels

Exercise Begins 30 Min. 45 Min. 1 Hr. 1 Hr. 30 Min.
Insulin levels drop. Glucagon levels rise. HGH levels rise.

Certain yoga postures, which have been used for thousands of years, help to accomplish these goals, especially improving blood flow to such digestive organs as the liver, pancreas, and gastrointestinal system. Therefore, the PGH Gene Makeover exercise program includes some of these basic time-tested and medically confirmed yoga postures.

The second type of essential activity is breathing. Although not normally considered an exercise, breathing can be extremely valuable in improving pulmonary functions and oxygen delivery to the brain and to the tissues. That helps remove excess acidity and helps balance not only the blood pH but also cellular fluids. Slow, deliberate deep breathing over time also produces a marked relaxation effect on the central nervous system by actually changing your brainwave patterns and leaving you in a calmer, more focused mental state. Whether breathing is performed in conjunction with yoga postures or on its own, the beneficial effects of focused breathing on many of the biomarkers of aging have been documented to be extremely effective.

SUMMARY OF PGH GENE MAKEOVER EXERCISE PROGRAM BENEFITS

Regular exercise has many beneficial effects on the reversible biomarkers of aging.

- Promotes fat burning.
- Optimizes hormonal levels.
- Improves muscle repair and muscle mass.
- Aids joint mobility.
- Enhances brain chemistry.
- Promotes nervous system balance and calming.
- Optimizes healthy cardiovascular function.
- Prevents disease.
- Cures some diseases.
- Improves appearance.

The PGH Gene Makeover anti-aging exercise program offers other benefits:

- Improves the way you feel and your outlook on life. Exercise releases so-called brain chemicals: epinephrine, norepinephrine, and painkilling endorphins. It also lowers the stress hormone cortisol. All of this leads to improved memory, better concentration, and a more positive attitude.

- Sculpts your body shape and improves your self-image.

- Enhances "executive functions" in the brain by increasing blood flow, causing the release of nitric oxide, a powerful brain stimulant.

- Improves vigor, with reduced levels of fatigue, confusion, and negative moods.

- Decreases anger and negative lipid profile, which significantly lowers the risk of cardiovascular heart disease.

- Relieves osteoarthritis pain.

SUGGESTED READING FOR EXERCISE PROGRAMS

The following books are excellent sources of the various types of recommended exercises. You will find many other helpful books at your library or bookstore. You might also consider taking yoga, tai chi, or qigong classes at your gym or a neighborhood recreation center.

Ansari, Mark, and Liz Lark. *Yoga for Beginners.* New York: HarperCollins, 1999.

Brzycki, Matt, and Fred Fornicola. *Dumbbell Training for Strength and Fitness.* Indianapolis, IN: Blue River Press, 2006.

Fenton, Mark. *Walking Magazine The Complete Guide To Walking: for Health, Fitness, and Weight Loss.* Guilford, CT: Lyons Press, 2001.

Horwitz, Tem, and Susan Kimmelman. *Tai Chi Ch'uan: The Technique of Power.* Chicago, IL: Cloud Hands, Inc. 2004.

Knopf, Karl. *Stretching for 50+: A Customized Program for Increasing Flexi-*

bility, Avoiding Injury, and Enjoying an Active Lifestyle. Berkeley, CA: Ulysses Press, 2004.

Knopf, Karl. *Weights for 50+: Building Strength, Staying Healthy, and Enjoying an Active Lifestyle.* Berkeley, CA: Ulysses Press, 2005.

McKenzie, Eleanor, Trevor Blount, and Joseph H. Pilates. *The Joseph H. Method at Home: A Balance, Shape, Strength, & Fitness Program.* Berkeley, CA: Ulysses Press, 2000.

Glossary

Allele Alternative form of a genetic locus (position); a single allele for each locus is inherited from each parent (for example, at the locus for eye color the allele might result in blue or brown eyes).

Allogeneic Variation in alleles among members of the same species.

Amino acid Any of a class of twenty molecules that are combined to form proteins in living things. The sequence of amino acids in a protein and hence protein function are determined by the genetic code.

Apoptosis Programmed cell death; the body's normal method of disposing of damaged, unwanted, or unneeded cells.

Base One of the molecules that form DNA and RNA molecules.

Base pair (bp) Two nitrogenous bases (adenine and thymine or guanine and cytosine) held together by weak bonds. Two strands of DNA are held together in the shape of a double helix by the bonds between base pairs.

Base sequence The order of nucleotide bases in a DNA molecule; determines structure of proteins encoded by that DNA.

Behavioral genetics The study of genes that may influence behavior.

Biotechnology A set of biological techniques developed through basic research and now applied to research and product development. In particular, biotechnology refers to the use by industry of recombinant DNA, cell fusion, and new bioprocessing techniques.

Birth defect Any harmful trait, physical or biochemical, that is present at birth, whether a result of a genetic mutation or some other nongenetic factor.

Cancer Diseases in which abnormal cells divide and grow unchecked. Cancer can spread from its original site to other parts of the body and can be fatal.

Carcinogen Something which causes cancer to occur by causing changes in a cell's DNA.

Cell The basic unit of any living organism that carries on the biochemical processes of life.

Centromere A specialized chromosome region to which spindle fibers attach during cell division.

Chromosomal deletion The loss of part of a chromosome's DNA.

Chromosome The self-replicating genetic structure of cells containing the cellular DNA that bears in its nucleotide sequence the linear array of genes. In prokaryotes, chromosomal DNA is circular, and the entire genome is carried on one chromosome. Eukaryotic genomes consist of a number of chromosomes whose DNA is associated with different kinds of proteins.

Clone An exact copy made of biological material such as a DNA segment (for example, a gene or other region), a whole cell, or a complete organism.

Congenital Any trait present at birth, whether the result of a genetic or nongenetic factor.

Crossing over The breaking during meiosis of one maternal and one paternal chromosome, the exchange of corresponding sections of DNA, and the rejoining of the chromosomes. This process can result in an exchange of alleles between chromosomes.

Cytosine (C) A nitrogenous base, one member of the base pair GC (guanine and cytosine) in DNA.

Deletion A loss of part of the DNA from a chromosome; which can lead to a disease or abnormality.

Deoxyribose A type of sugar that is one component of DNA (deoxyribonucleic acid).

Diploid A full set of genetic material consisting of paired chromosomes, one from each parental set. Most animal cells except the gametes have a diploid set of chromosomes. The diploid human genome has forty-six chromosomes.

Disease-associated genes Alleles carrying particular DNA sequences associated with the presence of disease.

DNA (deoxyribonucleic acid) The molecule that encodes genetic information. DNA is a double-stranded molecule held together by weak bonds between base pairs of nucleotides. The four nucleotides in DNA contain the bases adenine (A), guanine (G), cytosine (C), and thymine (T). In nature, base pairs form only between A and T and between G and C; thus the base sequence of each single strand can be deduced from that of its partner.

DNA repair genes Genes that encode proteins that correct errors in DNA sequencing.

DNA replication The use of existing DNA as a template for the synthesis of new DNA strands. In humans and other eukaryotes, replication occurs in the cell nucleus.

DNA sequence The relative order of base pairs, whether in a DNA fragment, gene, chromosome, or an entire genome.

Dominant An allele that is almost always expressed, even if only one copy is present.

Double helix The twisted-ladder shape that two linear strands of DNA assume when complementary nucleotides on opposing strands bond together.

Embryonic stem (ES) cells An embryonic cell that can replicate indefinitely, transform into other types of cells, and serve as a continuous source of new cells.

Enzyme A protein that acts as a catalyst, speeding the rate at which a biochemical reaction proceeds but not altering the direction or nature of the reaction.

Eukaryote Cell or organism with a membrane-bound, structurally discrete nucleus and other well-developed subcellular compartments. Eukaryotes include all organisms except viruses, bacteria, and blue-green algae.

Expression see Gene expression

Fraternal twin Siblings born at the same time as the result of fertilization of two ova by two sperm. They share the same genetic relationship to each other as any other siblings.

Functional genomics The study of genes, their resulting proteins, and the role played by the proteins in the body's biochemical processes.

Gamete Male or female reproductive cells; that is a sperm or an egg.

Gene The fundamental physical and functional unit of heredity. A gene is an ordered sequence of nucleotides located in a particular position on a particular chromosome that encodes a specific functional product (for example, a protein or RNA molecule).

Gene expression The process by which a gene's coded information is converted into the structures present and operating in the cell. Expressed genes include those that are transcribed into mRNA (messenger RNA) and then translated into protein and those that are transcribed into RNA but not translated into protein (for example, transfer and ribosomal RNAs).

Gene pool All the variations of genes in a species.

Gene therapy An experimental procedure aimed at replacing, manipulating, or supplementing nonfunctional or misfunctioning genes with healthy genes.

Genetic code The sequence of nucleotides, coded in triplets (codons) along the mRNA, that determines the sequence of amino acids in protein synthesis. A gene's DNA sequence can be used to predict the mRNA sequence, and the genetic code can in turn be used to predict the amino acid sequence.

Genetic illness Sickness, physical disability, or other disorder resulting from the inheritance of one or more deleterious alleles.

Genetic polymorphism Difference in DNA sequence among individuals, groups, or populations (for example, genes for blue eyes versus genes for brown eyes).

Genetic predisposition Susceptibility to a genetic disease. May or may not result in actual development of the disease.

Genetics The study of inheritance patterns of specific traits.

Genome All the genetic material in the chromosomes of a particular organism; its size is generally given as its total number of base pairs.

Genomics The study of genes and their function.

Genotype The genetic constitution of an organism, as distinguished from its physical appearance (its phenotype).

Guanine (G) A nitrogenous base, one member of the base pair GC (guanine and cytosine) in DNA.

Haploid A single set of chromosomes (half the full set of genetic material) present in the egg and sperm cells of animals and in the egg and pollen cells of plants. Human beings have twenty-three chromosomes in each of their reproductive cells, rather than the full set.

Hereditary cancer Cancer that occurs due to the inheritance of an altered gene within a family.

Homozygote An organism that has two identical alleles of a gene.

Hybrid The offspring of genetically different parents.

Identical twin Twins produced by the division of a single zygote; both have identical genotypes.

In vitro Studies performed outside a living organism, such as in a laboratory.

In vivo Studies carried out in living organisms.

Inherit In genetics, to receive genetic material from parents through biological processes.

Karyotype The chromosomal characteristics, or picture, of a cell.

Messenger RNA (mRNA) RNA that serves as a template for protein synthesis.

Mitochondrial DNA The genetic material found in mitochondria, the organelles that generate energy for the cell. Not inherited in the same fashion as nucleic DNA.

Mutagen An agent that causes a permanent genetic change in a cell. Does not include changes occurring during normal genetic recombination.

Mutation Any heritable change in DNA sequence.

Nitrogenous base A nitrogen-containing molecule having the chemical properties of a base. DNA contains the nitrogenous bases adenine (A), guanine (G), cytosine (C), and thymine (T).

Nucleic acid A large molecule composed of nucleotide subunits.

Nucleotide A subunit of DNA or RNA consisting of a nitrogenous base (adenine, guanine, thymine, or cytosine in DNA; adenine, guanine, uracil, or cytosine in RNA), a phosphate molecule, and a sugar molecule (deoxyribose in DNA and ribose in RNA). Thousands of nucleotides are linked to form a DNA or RNA molecule.

Nucleus The cellular organelle in eukaryotes that contains most of the genetic material.

Oncogene A gene, one or more forms of which are associated with cancer. Many oncogenes are involved, directly or indirectly, in controlling the rate of cell growth.

Pharmacogenomics The study of the interaction of an individual's genetic makeup and response to a drug.

Phenotype The physical characteristics of an organism or the presence of a disease that may or may not be genetic.

Polymorphism Difference in DNA sequence among individuals that may underlie differences in health. Genetic variations occurring in more than 1 percent of a population would be considered useful polymorphisms for genetic linkage analysis.

Protein A large molecule composed of one or more chains of amino acids in a specific order; the order is determined by the base sequence of nucleotides in the gene that codes for the protein. Proteins are required for the structure, function, and regulation of the body's cells, tissues, and organs; and each protein has unique functions. Examples are hormones, enzymes, and antibodies.

Pyrimidine A nitrogen-containing single-ring basic compound that occurs in nucleic acids. The pyrimidines in DNA are cytosine and thymine; in RNA, cytosine and uracil.

Recessive gene A gene which will be expressed only if there are two identical copies or, for a male, if one copy is present on the X chromosome.

Ribose The five-carbon sugar that serves as a component of RNA.

Ribosomal RNA (rRNA) A class of RNA found in the ribosomes of cells.

RNA (Ribonucleic acid) A chemical found in the nucleus and cytoplasm of cells; it plays an important role in protein synthesis and other chemical activities of the cell. The structure of RNA is similar to that of DNA. There are several classes of RNA molecules, including messenger RNA, transfer RNA, ribosomal RNA, and other small RNAs, each serving a different purpose.

Sex chromosome The X or Y chromosome in human beings that determines the sex of an individual. Females have two X chromosomes in diploid cells; males have an X and a Y chromosome. The sex chromosomes comprise the twenty-third chromosome pair in a karyotype.

Single nucleotide polymorphism (SNP) DNA sequence variations that occur when a single nucleotide (A, T, C, or G) in the genome sequence is altered.

Somatic cell Any cell in the body except gametes and their precursors.

Stem cell Undifferentiated, primitive cells in the bone marrow that have the ability both to multiply and to differentiate into specific blood cells.

Syndrome The group or recognizable pattern of symptoms or abnormalities that indicate a particular trait or disease.

Telomerase The enzyme that directs the replication of telomeres.

Telomere The structures at the two ends of a chromosome. This specialized structure is involved in the replication and stability of linear DNA molecules.

Teratogenic Substances such as chemicals or radiation that cause abnormal development of a embryo.

Thymine (T) A nitrogenous base, one member of the base pair AT (adenine-thymine).

Transcription The synthesis of an RNA copy from a sequence of DNA (a gene); the first step in gene expression.

Transfer RNA (tRNA) A class of RNA having structures with triplet nucleotide sequences that are complementary to the triplet nucleotide coding sequences of mRNA. The role of tRNAs in protein synthesis is to bond with amino acids and transfer them to the ribosomes, where proteins are assembled according to the genetic code carried by mRNA.

Translation The process in which the genetic code carried by mRNA directs the synthesis of proteins from amino acids.

Uracil A nitrogenous base normally found in RNA but not DNA; uracil is capable of forming a base pair with adenine.

Wild type The form of an organism that occurs most frequently in nature.

X chromosome One of the two sex-determining chromosomes, X and Y. The presence of two X chromosomes in the pair is the determinant for a female.

Y chromosome One of the two sex-determining chromosomes, X and Y. The presence of an X chromosome in the pair is the determinant for a male.

Resources

This section contains resources of a variety of anti-aging products, medical personnel, books, journals, and newsletters.

Anti-Aging Clinic

Giampapa Institute
89 Valley Road
Montclair, NJ 07042
Phone: 973-746-3535
www.skinbodyclinic.com

Anti-Aging Products

Suracell
Personal Genetic Health
87 Valley Road
Montclair, NJ 07042
Phone: 973-932-1200
www.suracell.com

Maximum Human Performance
21 Dwight Place
Fairfield, NJ 07004
Phone: 888-783-8844
www.maxperformance.com/
home.php

Home Test Kits

Saliva Hormone Tests
Anti-Aging Hormone Profile Kit
Life-Flo Health Care Products
11202 North 24th Avenue
Phoenix, AZ 85029-4745
www.life-flo.com

Serum Thiol Test for DNA Repair
Art Banne, Lab Director
Biomedical Diagnostic Research, LLC
8140 Mayfield Road
Chesterland, OH 44026
Phone: 440-729-6080

Tests for Your Doctor

Suracell
Personal Genetic Health
87 Valley Road
Montclair, NJ 07042
Phone: 973-932-1200
www.suracell.com

F_2–Isoprostanes Urine or Serum Immunoassay

Oxford Biomedical Research, Inc.
P.O. Box 522
Oxford, MI 48371
Phone: 248-628-5104
   or 800-692-4633
www.oxfordbiomed.com

Saliva Home Assay

Hormone View Personal Hormone
   Profile
Ewald Pretner, M.D., Medical
   Director
AllVia Diagnostic Laboratories,
   LLC
11202 North 24th Avenue
Phoenix, AZ 85029-4745

Serum Protein Thiol Test

Art Banne, Lab Director
Biomedical Diagnostic Research,
   LLC
8140 Mayfield Road
Chesterland, OH 44026
Phone: 440-729-6080

Spectra Cell Laboratories, Inc.

(specializes in cell micronutrient
   analysis)
7051 Portwest Drive, Suite 100
Houston, TX 77024-8026
Phone: 800-227-5227
www.spectracell.com

Compounding Pharmacies (Require prescription)

Hopewell Pharmacy

1 West Broad Street
Hopewell, NJ 08525
www.hopewellrx.com

Profile Health, Medical Consultations

Debra Lassiter, R.N., L.N.P.
Phoenix, AZ 85377
Phone: 480-488-2007

Women's International Pharmacy

Natural Hormone Therapy
2 Marsh Court
Madison, WI 53718
Phone: 800-279-5708
www.womensinternational.com
or
12012 North 111th Avenue
Youngtown, AZ 85363
Phone: 800-279-5708

Transdermal Hormone Creams

Life-Flo Health Products
11202 North 24th Avenue
Phoenix, AZ 85029-4745
Phone: 888-999-7440
www.life-flo.com

University Compounding Pharmacy

1875 3rd Avenue
San Diego, CA 92101
Phone: 800-985-8065

Locating an Anti-aging Physician

Alliance for Aging Research
2021 K Street, NW, Suite 305
Washington, DC 20006
Phone: 202-293-2856
Fax: 202-785-8574
www.agingresearch.org

American Academy of Aesthetic Medicine (AAAM)
645-375 Water Street
Vancouver, BC V6B 5C6
Canada
Phone: 1-604-681-5226
Fax: 1-604-681-2503

American Academy of Anti-Aging Medicine (A4M)
401 North Michigan Avenue
Chicago, IL 60611-4267
Phone: 312-321-6869
www.aaamed.org

American College of Advancement in Medicine (ACAM)
23121 Verdugo Drive
Laguna Hills, CA 92653
Phone: 714-583-7666
Fax: 714-4555-9679
www.acamnet.org

American Aging Association
2129 Providence Avenue
Chester, PA 19013
Phone: 610-874-7550
Fax: 610-876-7715
www.americanaging.org

Canadian Association of Aesthetic Medicine (CAAM)
1179 Eastview Road
North Vancouver, BC V7J1L7
Phone: 604-805-1897
Fax: 604-984-2530
www.caam.ca

Harvard Medical School
Division on Aging
643 Huntington Avenue
Boston, MA 02115
Phone 617-432-1840
http://positiveaging.org/hms.html

International Academy of Alternative Health and Medicine
218 Avenue B
Redondo Beach, CA 90277
Phone: 310-540-0564
Fax: 310-540-0564

International Federation on Aging
601 E Street, NW
Washington, DC 20049
Phone: 202-434-2427
Fax: 202-434-6458

New York Academy of Sciences
2 East 63rd Street
New York, NY 10021
Phone: 212-838-0230
Fax: 212-888-2894
www.nyas.org

Books

Ackland, Lesley. *Pilates Over 50.* London: Hammersmith, 2001.

Bell, Lorna. *Gentle Yoga.* Berkeley, CA: Celestial Arts, 2000.

Carrico, Mara. *Yoga Basics.* New York: Henry Holt, 1997.

Cheum, Lam Kam. *Tai Chi.* London: Gaia Books, 1994.

Frost, Simon. *Stretching.* New York: Sterling Publishing, 2002.

Francina, Suza. *The New Yoga for People Over 50.* Deerfield Beach, FL: Health Communications, 1997.

Giampapa, Vincent, and Ronald Pero. *The Anti-Aging Solution: 5 Simple Steps to Looking and Feeling Young.* New York: John Wiley & Sons, 2004.

Giampapa, Vincent. *Basic Principles: Anti-Aging Medicine and Age Management.* Montclair, NJ: Giampapa Institute, 2003. (Available from Giampapa Institute.)

Lin, Chunyi. *Born a Healer.* Minneapolis, MN: Spring Forest Qigong. www.mnwelldir.org/docs/qigong/qigong3.htm.

Miller, Emmett. *Deep Healing,* Carlsbad, CA: Hay House, 1997.

Page, Christine. *Beyond the Obvious.* Essex, UK: Daniel Company, Limited, 1998.

Perl, Candace B. *Molecules of Emotion.* New York: Touchstone, 1997.

Pero, Ronald, and Marcia Zimmerman. *Reverse Aging Through the Miracle of Natural DNA Repair.* Nutrition Solution Press, 2002.

Roberts, Matt. *90-Day Fitness Plan.* New York: DK Publishing, 2001.

Searle, Sally, and Cathy Meeus. *Secrets of Pilates.* New York: DK Publishing, 2001.

Wolf, Fred Alan. *The Spiritual Universe: One Physicist's Vision of Spirit, Soul, Matter, and Self.* Portsmouth, NH: Moment Point Press, 1996.

Yuan, Chun-Su, and Eric J. Bieber. *Textbook of Complementary and Alternative Medicine,* Boca Raton, FL: CRC Press, 2003.

Zimmerman, Marcia. *Eat Your Colors—Maximize Your Health by Eating the Right Foods for Your Body Type.* New York: Henry Holt, 2002.

Journals

Journal of Anti-Aging Medicine
Mary Ann Liebert, Inc., Publishers
2 Madison Avenue
Larchmont, NY 10538
Phone: 914-834-3689
www.grg.org/resources/jaam.html

Townsend Letter (previous title: *Townsend Letter for Doctors and Patients*)
911 Tyler Street
Port Townsend, WA 98368
Phone: 360-385-0699
www.townsendletter.com

Newsletters

Environmental Nutrition
52 Riverside Drive
New York, NY 10024-6599
www.environmentalnutrition.com/

Focus on Healthy Aging
Mt. Sinai School of Medicine
Box 420235
Palm Coast, FL 32142-0235
Phone: 800-829-9406
www.focusonhealthyaging.com

Giampapa Institute Newsletter
89 Valley Road
Montclair, NJ 07042
Phone: 973-746-3535

Harvard Health Letter and *Harvard Women's Health Watch*
10 Shattuck Street, Suite 612
Boston. MA 02115
www.health.harvard.edu

Nutrition News
Riverside, CA
Free to customers of nutritional stores throughout the United States

The Nutrition Reporter
P.O. Box 30246
Tucson, AZ 85751-0246
www.thenutritionreporter.com

Taste for Life
Peterborough, NH
Free to customers of nutrition markets throughout the United States
www.tasteforlife.com

Physician Educational Seminar Series

contact: **Giampapa Institute**
            89 Valley Road
            Montclair, NJ 07042
            Phone: 973-746-3535

References

This abridged section contains a sample of the scientific discoveries and evidence supporting the Personal Genetic Health approach. A complete listing of the research citations is located at www.thegenemakeover.com/references.

Chapter 1. Proof We Can Slow Aging Now

Bodkin, N.L., HK Ortmeyer, and BC Hansen. "Long-term dietary restriction in older-aged rhesus monkeys: effects on insulin resistance." *The Journals of Gerontology, Series A: Biological Sciences and Medical Science* 50:3 (May 1995): B$_1$42–147.

Cao, Shelly X., et al. "Genomic Profiling of Short- and Long-Term Calorie Restriction in the Liver of Aging Mice." *Proceedings of the National Academy of Sciences* 98 (2001): 10630–10635.

Chan, Y.C., et al. "Dietary, anthropometic, hematological and biochemical assessment of the nutritional status of centenarians and elderly people in Okinawa, Japan." *Journal of the American College of Nutrition* 16:3 (1997): 229–235.

Chung, H.Y., et al. "Molecular inflammation hypothesis of aging based on the antiaging mechanism of calorie restriction." *Microscopy Research and Technique* 59 (2002): 264–272.

Corton, J.C., and H.M. Brown-Borg. "Peroxisome proliferator-activated receptor coactivator 1 in calorie restriction and other models of longevity."

The Journals of Gerontology Series A: Biological Sciences and Medical Sciences 60 (2005): 1494–1509.

Dhahbi, J.M., and S.R. Spindler, "Aging of the Liver," *Biology of Aging and its Modulation: Aging of the Organs and Systems,* Vol. 3, ed. R. Aspinall, (Dordecht, The Netherlands: Kluwer Academic Publishers, 2004), 271–291.

Dhahbi, J.M., et al. "Caloric intake alters the efficiency of catalase mRNA translation in the liver of old female mice." *The Journals of Gerontology, Series A: Biological Sciences and Medical Science.* 53 (1998): B$_1$80–185.

Dhahbi, J.M., et al. "Caloric restriction alters the feeding response of key metabolic enzyme genes." *Mechanisms of Ageing and Development* 122 (2001): 1033–1048.

Dhahbi, J.M., et al. "Calories and aging alter gene expression for gluconeogenic, glycolytic, and nitrogen-metabolizing enzymes." *American Journal of Physiology* 277 (1999): 352–360.

Dhahbi, J.M., et al. "Chaperone-mediated regulation of hepatic protein secretion by caloric restriction." *Biochemical and Biophysical Research Communications* 284 (2001); 335–339.

Dhahbi, J.M., et al. "Dietary energy tissue-specifically regulates endoplasmic reticulum chaperone gene expression in the liver of mice." *Journal of Nutrition* 127 (1997): 1758–1764.

Dhahbi, J.M., et al. "Hepatic gene expression profiling of streptozotocin-induced diabetes." *Diabetes Techniques and Therapeutics* 5 (2003): 411–420.

Dhahbi, J.M., et al. "Postprandial induction of chaperone gene expression is rapid in mice." *Journal of Nutrition* 132 (2002): 31–37.

Dhahbi, J.M., et al. "Temporal Linkage Between the Phenotypic and Genomic Responses to Caloric Restriction." *Proceedings of the National Academy of Sciences* 101:15 (Apr 2004): 5524–5529.

Fontana, L., et al. "Long-term calorie restriction is highly effective in reducing the risk for atheroscerosis in humans." *Proceedings of the National Academy of Sciences* 101 (2004): 6659–6663.

Gesteland, R.F., and J.F. Atkins, eds. *The RNA World.* Cold Spring Harbor, NY: Cold Spring Harbor Laboratory Press; 1993.

Grube, K., and A. Burkle. "Poly(ADP-ribose) polymerase activity in mononuclear leukocytes of 13 mammalian species correlates with species-specific life span." *Proceedings of the National Academy of Sciences* 89 (1992): 11759–11763.

Hammarqvist, F., et al. "Free amino acid and glutathione concentrations in muscle during short-term starvation and refeeding." *Clinical Nutrition* 24:2 (April 2005): 236–243.

Heilbronn, L.K., and E. Ravussin. "Calorie restriction and aging: review of the literature and implications for studies in humans." *American Journal of Clinical Nutrition* 78 (2003): 361–369.

Heilbronn, L.K., et al. "Effect of 6–month calorie restriction on biomarkers of longevity, metabolic adaptation, and oxidative stress in overweight individuals." *JAMA, the Journal of the American Medical Association* 295 (2006): 1539–1548.

Jeffares, D.C., A.M. Poole, and D. Penny. "Relics from the RNA world." *Journal of Molecular Evolution* 46 (1998): 18–36.

Kemnitz, J.W., et al. "Dietary restriction of adult male rhesus monkeys: design, methodology, and preliminary findings from the first year of study." *Journal of Gerontology* 48:1 (Jan 1993): B17–26.

Kim, H.J., et al. "Modulation of redox-sensitive transcription factors by calorie restriction during aging." *Mechanisms of Ageing and Development* 123 (2002): 1589–1595.

Lamm, S., Y. Sheng, Y., and R.W. Pero. "Persistent response to pneumococcal vaccine in individuals supplemented with a novel water soluble extract of *Uncaria tomentosa*, C-MED-100." *Phytomedicine* 8:4 (2001): 267–274.

Lane, M.A., et al. "Calorie restriction in nonhuman primates: effects on diabetes and cardiovascular disease risk." *Toxicological Sciences* 52 (1999): 41–48.

Lane, M.A., et al. "Dehydroepiandrosterone sulfate: a biomarker of primate aging slowed by calorie restriction." *Journal of Clinical Endocrinology & Metabolism* 82 (1997): 2093–2096.

Lee, C.M., R. Weindruch, and J.M. Aiken. "Age associated alterations of the mitochondrial genome." *Free Radical Biology and Medicine* 22:7 (1997): 1259–1269.

Lee, J., et al. "Dietary restriction increases the number of newly generated neural cells, and induces BDNF expression, in the dentate gyrus of rats." *Journal of Molecular Neuroscience* 15 (2000): 99–108.

Lodish, H. *Molecular Cell Biology,* Chapter 12. New York: WH Freeman and Co, 2000.

Lopez-Lluch, G., et al. "Calorie restriction induces mitochondrial biogenesis and bioenergetic efficiency." *Proceedings of the National Academy of Sciences* 103 (2006): 1768–1773.

McCay C.M., M.F. Crowell, and L.A. Maynard. "The effect of retarded growth upon the length of lifespan and upon the ultimate body size." *Journal of Nutrition* 10 (1935): 63–79.

Chapter 2. The First Secret Revealed

Ballou, S.P., et al. "Quantitative and qualitative alterations of acute-phase proteins in healthy elderly persons." *Age and Aging* 25 (1996): 224–230.

Blackburn, E., et al. "Recognition and Elongation of Telomeres by Telomerase" *Genome* 312 (1989): 553–560.

Broussolle, C., et al. "Evaluation of the fructosamine test in obesity: Consequences for the assessment of past glycemic control in diabetes." *Clinical Biochemistry* 24 (1991); 203–209.

Busse, E., et al. "Influence of a-lipoic acid on intra-cellular glutathione in vitro and in vivo." *Arzneimittelforschung* 42:6 (1992): 829–831.

Cabana, V.G., J.N. Siegel, and Sabesin SM. "Effects of the acute phase response on the concentration and density distribution of plasma lipids and apolipoproteins." *Journal of Lipid Research,* 30 (1989):30: 39–49.

Chandra, R.K. "Nutrition and the immune system: an introduction." *American Journal of Clinical Nutrition* 66:2 (1997): 460S-463S.

Clancy, R.M., and S.B. Abramson. "Nitric oxide: A novel mediator of inflammation." *Proceedings of the Society for Experimental Biology and Medicine* 210:2 (1995): 93–101.

Cooper, G.J., and C.A. Tse. "Amylin, amyloid and age-related disease." *Drugs & Aging* 9:3 (1996): 202–212.

Coufturier, M., et al. "Variable glycation of serum proteins in patients with diabetes mellitus." *CIM: Clinical and Investigative Medicine* 20:2 (1997): 103–109.

Coussons, P.J., et al. "Glucose modification of human serum albumin: A structural study." *Free Radical Biology and Medicine*;22:7 (1997): 1217–1227.

Dandona, P., et al. "The suppressive effect of dietary restriction and weight loss in the obese on the generation of reactive oxygen species by leukocytes, lipid peroxidation, and protein carbonylation." *Journal of Clinical Endocrinology & Metabolism* 86 (2001): 355–362.

Dills, W.L. "Protein fructosylation: Fructose and the Maillard reaction." *American Journal of Clinical Nutrition*;58:Suppl (1993): 779S-787S.

Droge, W. "Oxidative aging and insulin receptor signaling." *The Journals of Gerontology, Series A: Biological Sciences and Medical Science* 60:11 (Nov 2005): 1378–1385.

Finch, C.E.. "A perspective on sporadic inclusion-body myositis." *Neurology* 66 (2006): S1–S6.

Firestein, G.S., and N.J. Zvaifler. "Anticytokine therapy in rheumatoid arthritis." *New England Journal of Medicine* 337:3 (1997): 195–197.

Fleming J.E., et al. "Age-dependent changes in proteins *of Drosofhila melanogaster.*" *Science.* 231 (1986): 1157–1159.

Giugliano, D., A. Ceriello, and G. Paolisso. "Diabetes mellitus, hypertension, and cardiovascular disease: which role for oxidative stress?" *Metabolism.* 44:3 (1995): 363–368.

Goodman, J. "Histone tails wag the DNA dog." *Helix.* [University of VA Health System]. 17: 1 (Spring 2000).

Gredilla, R., and G. Barja. "Minireview: the role of oxidative stress in relation to calorie restriction and longevity." *Endocrinology* 146:9 (2005): 3713–3717.

Gresl, T.A., et al. "Dietary restriction and glucose regulation in aging rhesus monkeys: a follow-up report at 8.5 yr." *American Journal of Physiology – Endocrinology and Metabolism* 281:4 (Oct 2001): E757–765.

Guillasseau, P.J., et al. "Comparison of fructosamine with glycated hemoglobin as an index of glycemic control in diabetic patients." *Diabetes Research* 13 (1990): 127–131.

Hamilton, M.L., et al. "Does oxidative damage to DNA increase with age?" *Proceedings of the National Academy of Sciences* 98:18 (Aug 2001): 10469–10474.

Hasdai, D., et al. "Increased serum concentrations of interleukin-lb in patients with coronary artery disease." *Heart*;76:l (1996): 24–28.

Haverkate, F., et al. "Production of C-reactive protein and risk of coronary events in stable and unstable angina." *Lancet* 349:9050 (1997): 462–466.

Hilliquin, P. "Biological markers in inflammatory rheumatic diseases." *Cellular and Molecular Biology* 4l:8 (1995): 993–1006.

Jonas, W.B., C.P. Rapoza, and W.E. Blair. "The effect of niacinamide on osteoarthritis: pilot study." *Inflammation Research* 45:l (1996): 330–334.

Kaufman, W. "Niacinamide therapy for joint mobility: therapeutic reversal of a common clinical manifestation of the normal aging process." *Connecticut State Medical Journal* 17 (1953) 584–589.

Khan, S., and J. Rupp. "The effect of exercise conditioning, diet and drug therapy on glycosylated hemoglobin levels in type 2 (NIDDM) diabetics." *Journal of Sports Medicine and Physical Fitness* 35 (1995): 281–288.

Kimura, K.D., et al. "daf-2, An insulin receptor like gene that regulates longevity and diapause in *Caenorhabditis elegans*." *Science* 277:5328 (1997): 942–946.

Kimura, T., et al. "Identification of advanced glycation end products of the Maillard reaction in Pick's disease." *Neuroscience Letters* 219 (1996): 95–98.

King, G.L., and M. Brownlee. "The cellular and molecular mechanisms of diabetic complications." *Endocrinology and Metabolism Clinics North America* 25:2 (1996): 255–270.

Knecht, K.J., et al. "Effect of diabetes and aging on carboxymethyllysine levels in human urine." *Diabetes* 40 (1991): 190–196.

Lapolla, A., et al. "Glycosylated serum proteins in diabetic patients and their relation to metabolic parameters." *Diabète Métabolisme* 11 (1985): 238–242.

Meloni, T., et al. "HbAlc levels in diabetic Sardinian patients with or without G6PD deficiency." *Diabetes Research and Clinical Practice* 23:1 (1994): 59–61.

Mendall, M.A., et al. "C-reactive protein and its relation to cardiovascular risk factors: A population based cross sectional study." *British Medical Journal* 312:1038 (1996): 1061–1065.

Miesel, R., M. Kurpisz, and H. Kroger. "Modulation of inflammatory arthritis by inhibition of poly(ADP ribose) polymerase." *Inflammation* 19:3 (1995): 379–387.

Miyata, T., et al. "The receptor for advanced glycation end products (RAGE) is a central mediator of the interaction of AGE-p2microglobulin with human mononuclear phagocytes via an oxidant-sensitive pathway." *Journal of Clinical Investigation* 98:5 (1996): 1088–1094.

Moulton, P.J. "Inflammatory joint disease: the role of cytokines, cyclooxygenases and reactive oxygen species." *British Journal of Biomedical Science* 53:4 (1996): 317–324.

National Institutes of Health Press Release. *Researchers identify gene for premature aging disorder: Progeria gene discovery may help solve mysteries of normal aging.* (April 16, 2003) www.genome.gov/11006962.

Wright, W.E., and J.W. Shay. "Telomere biology in aging and cancer." *Journal of the American Geriatrics Society* 53:9 Suppl (Sep 2005): S292–4.

Chapter 3. The Second Secret Revealed

Abbasoglu, O., et al. "The effect of the pineal gland on liver regeneration in rats." *Journal of Hepatology* 23 (1995): 578–581.

Alpini, D., et al. "Aging and vestibular system: specific tests and role of melatonin in cognitive involvement." *Archives of Gerontology and Geriatrics* Suppl. 9 (2004): 13–25.

Ardinali, D.P., et al. "Melatonin effects on bone: experimental facts and clinical perspectives." *Journal of Pineal Research* 34 (2003): 81–87.

Beach, R.L., et al. "The identification of neurotrophic factor as a transferrin." FEBS Letters 156 (1983): 151–156.

Bhatt, R., et al. "Effects of kindled seizures upon hematopoiesis in rats." *Epilepsy Research* 54 (2003): 209–219.

Bondy, S.C., et al. "Retardation of brain aging by chronic treatment with melatonin." *Annals of the New York Academy of Sciences* 1035 (2004): 197–215.

Bruinink, A., et al. "Neurotrophic effects of transferrin on embryonic chick brain and neural retinal cell cultures." *International Journal of Developmental Neuroscience* 14 (1996): 785–795.

Bubenik, G.A., "Gastrointestinal melatonin: localization, function, and clinical relevance." *Digestive Diseases and Sciences.* 47 (2002): 2336–2348.

Conboy, I.M., et al. "Rejuvenation of aged progenitor cells by exposure to a young systemic environment." *Nature* 433 (2005): 780–784.

Conti, A., et al. "Evidence for melatonin synthesis in mouse and human bone marrow cells." *Journal of Pineal Research* 28 (2000): 193–202.

Csaba, G. "Presence in and effects of pineal indoleamines at very low level of phylogeny." *Experientia* 49 (1993): 627–634.

Danilova, N., et al. "Melatonin stimulates cell proliferation in zebrafish embryo and accelerates its development." *The FASEB Journal* 18 (2004): 751–753.

Del Rio-Tsonis, K., and P.A. Tsonis. "Amphibian tissue regeneration: a model for cancer regulation" (review). *International Journal of Oncology* 1 (1992): 161–164.

Eguchi, G., and K. Watanabe. 1973. Elicitation of lens formation from the "ventral iris" epithelium of the newt by a carcinogen, N-methyl-N'-nitro-N-nitrosoguanidine. J. Embryol. Exp. Morphol. 30: 63–71.

Elias, A.N., et al. "Ketosis with enhanced GABAergic tone promotes physiological changes in transcendental meditation." *Medical Hypotheses* 54 (2000): 660–662.

Esrefoglu, M., et al. "Potent therapeutic effect of melatonin on aging skin in pinealectomized rats." *Journal of Pineal Research* 39 (2005): 231–237.

Fernandez, M.A., et al. "Intracellular trafficking during liver regeneration. Alterations in late endocytic and transcytotic pathways." *Journal of Hepatology* 40 (2004): 132–139.

Fischer, T.W., et al. "Melatonin increases anagen hair rate in women with androgenetic alopecia or diffuse alopecia: results of a pilot randomized controlled trial." *British Journal of Dermatology* 150 (2004): 341–345.

Forbes, S.J., et al. "Adult stem cell plasticity: new pathways of tissue regeneration become visible." *Clinical Science* (London) 103 (2002): 355–369.

Frangioni, G., et al. "Melatonin, melanogenesis, and hypoxic stress in the newt, *Triturus carnifex*." *Journal of Experimental Zoology* 296 (2003): 125–136.

Frank, L.A., et al. "Adrenal steroid hormone concentrations in dogs with hair cycle arrest (AlopeciaX) before and during treatment with melatonin and miotane." *Veterinary Dermatology* 15 (2004): 278–284.

Ghosh, A.K., 2002. "Factors involved in the regulation of type I collagen gene expression: implication in fibrosis." *Experimental Biology and Medicine* 227 (2002): 301–314.

Goolsby, J., et al. "Hematopoietic progenitors express neural genes." *Proceedings of the National Academy of Sciences* 100 (2003): 14926–14931.

Harty, M., et al. "Regeneration or scarring: an immunological perspective." *Developmental Dynamics.* 226 (2003): 268–279.

Haus, E., et al. "Stimulation of the secretion of dehydroepiandrosterone by melatonin in mouse adrenals *in vitro*." *Life Science* 58 (1996): 263–267.

Huang, J.S., et al. "Synthetic TGF-beta antagonist accelerates wound healing and reduces scarring." *FASEB Journal* 16 (2002): 1269–1270.

Kobayashi, H., et al. "A role of melatonin in neuroectodermal-mesodermal interactions: the hair follicle synthesizes melatonin and expresses functional melatonin receptors." *FASEB Journal* 19 (2005): 1710–1712.

Komori, M., et al. "Involvement of bone marrow-derived cells in healing of experimental colitis in rats." *Wound Repair Regen.* 13 (2005): 109–118.

Krause, D.S., "Plasticity of marrow-derived stem cells." *Gene Therapy.* 9 (2002): 754–758.

Kucia, M., et al. "Tissue-specific muscle, neural and liver stem/progenitor cells reside in the bone marrow, respond to an SDF-1 gradient and are mobilized into peripheral blood during stress and tissue injury." *Blood Cells, Molecules, and Diseases* 32 (2004): 52–57.

Kunz, D., and F. Bes. "Exogenous melatonin in periodic limb movement disorder: an open clinical trial and a hypothesis." *Sleep.* 24 (2001): 183–187.

Leferovich, J.M., et al. "Heart regeneration in adult MRL mice." *Proceedings of the National Academy of* Sciences 97 (2001): 2830–2835.

Leng, S., et al. "Serum interleukin-6 and hemoglobin as physiological correlates in the geriatric syndrome of frailty: a pilot study." *Journal of the American Geriatrics Society* 50 (2002): 1268– 1271.

Lesnikov, V.A., and W. Pierpaoli. "Pineal cross-transplantation (old-to-young and vice versa) as evidence for an endogenous 'aging clock'." *Annals of the New York Academy of Sciences* 719 (1994): 456–460.

Lesnikova, M., et al. "Upregulation of interleukin-10 and inhibition of alloantigen responses by transferrin and transferrin-derived glycans." *Journal of Hematotherapy* & *Stem Cell Research.* 9 (2000): 381–392.

Lio, D., et al. "Association between the MHC class I gene HFE polymorphism and longevity: a study in a Sicilian population." *Genes and Immunity.* 3 (2002): 20–24.

Lio, D., et al., "Inflammation, genetics, and longevity: further studies on the protective effects in men of IL-10–1082 promoter SNP and its interaction with TNFgamma-308 promoter SNP." *Journal of Medical Genetics* 40 (2003): 296–299.

Sheng, Y., et al. *Phytomedicine* 8:4 (August 2001). 275–282.

Chapter 4. The Third Secret Revealed

Alpha Tocopherol, Beta Carotene, Cancer Prevention Study Group. "The effect of vitamin E and beta carotene on incidences of lung cancer and other cancers in male smokers." *New England Journal of Medicine* 330:15 (April 1994): 1029–1035.

Arsenian, M.A., "Magnesium and cardiovascular disease." *Progress in Cardiovascular Diseases* 35 (1993): 271–310.

Ascherio, A., C.H. Hennekens, and W.C. Willett. "Trans-fatty acid intake and risk of myocardial infarction." *Circulation.* 89 (1994): 94–101.

Baggio, E., et al. "Italian multicenter study on the safety and efficacy of coenzyme QIO as adjunctive therapy in heart failure." *Molecular Aspects of Medicine* 15 (1994) 5287–5294.

Block, G., B. Patterson, and A. Sufar. "Fruit, vegetables and cancer prevention: a review of the epidemiological evidence." *Nutrition and Cancer* 18 (1992): l-29.

Blot, W.J., et al. "Nutritional intervention trials in Linxion, China." *Journal of National Cancer Research* 85 (1993):1483–1492.

Colditz, G.A., et al. "Increased green and leafy vegetable intake and lowered cancer deaths in an elderly population." *American Journal of Clinical Nutrition* 41 (1985): 32–36.

Dean, W. *The Biological Aging Measurement—Clinical Applications.* Pensacola, FL: Center for Bio-Gerontology, 1988.

Dhahbi, J.M., et al. "Postprandial Induction of chaperone gene expression is rapid in mice." *Journal of Nutrition.* 132 (2002): 31–37.

Evans, W., and I.H. Rosenberg. *Biomarkers.* New York: Simon & Schuster, 1991.

Giampapa, V.C., R. Klatz, and R. Goldman. "Anti-Aging Surgery: A Step Beyond Cosmetic Surgery." *Advances in Anti-Aging Medicine.* Vol. 1. New York: Mary Ann Liebert Publishers, 1996: 57–60.

Hayflick, L., *How and Why We Age.* New York: Ballantine Books, 1994.

Hill, E.G., et al. "Perturbation of the metabolism of essential fatty acids by dietary partially hydrogenated vegetable oil." *Proceedings of the National Academy of Sciences* 79 (1982): 953–957.

Lindheim, S.R., et al. "A possible bimodal effect of estrogen on insulin sensitivity in postmenopausal women and the attenuation effect of added progestin." *Fertility and Sterility* 60 (1993): 664–667.

Maurer, K., et al. "Clinical efficacy of gingko biloba special extract EGb 761 in dementia of Alzheimer type." *Journal of Psychiatric Research* 1997:31: 645–655.

Mohr, D., V.W. Bowry, and R. Stocker. "Dietary supplementation with coenzyme QlO results in increased levels of ubiquinol-10 within circulating lipoproteins and increased resistance of human low-density lipoproteins to the initiation of lipid peroxidation." *Biochimica et Biophysica Acta* 1126 (1992): 247–254.

Morley, J.E., et al. "Potentially predictive and manipulable blood serum correlates of aging in the healthy human male." *Proceedings of the National Academy of Sciences* 94 (1997): 7537–7542.

Murray, M.T. *Encyclopedia of Nutritional Supplements.* Rocklin, CA: Prima Publishing, 1996.

Oilman, V., and W. Dean. *Neuroendocrine Theory of Aging.* Pensacola, FL: Center for Bio-Gerontology, 1992.

Older and Wiser: The Baltimore Longitudinal Study of Aging. [NIH Publication No. 89–2797.] Washington, DC: U.S. Government Priming Office, 1989.

Polyp Prevention Group. "A clinical trial of antioxidant vitamins to prevent colorectal adenoma." *New England Journal of Medicine* 331 (1994): 141–147.

Rimm, E.B., et al. "Vitamin E consumption and risk of coronary heart disease in men." *New England Journal of Medicine* 328 (1993): 1450–1456.

Roberts, H.J. *Aspartame: Is It Safe?* Philadelphia, PA: Charles Press, 1990.

Shekelle, R.B., M. Lepper, and S. Liu. "Dietary vitamin A and risk of cancer in the Western Electric Study." *Lancet* 2 (1981): 1185–1190.

Shils, M.E., and J.A. Olson. *Modern Nutrition in Health and Diseases,* 9th ed. Baltimore, MD: Lippincott Williams & Wilkins, 1999: 573–603.

Stampfer, M.J., et al. "Vitamin E consumption and risk of coronary disease in women." *New England Journal of Medicine* 32 (1993): 1444–1449.

Steinmetz, K.A., and J.C. Potter. "Vegetables, fruit and cancer. I. Epidemiology." *Cancer Causes & Control* 2:5 (1991): 325–357.

The Duke Longitudinal Studies of Normal Aging 1955–1980: An Overview of History, Design, and Findings. New York: Springer Publishing Co., 1985.

Timiras, P.S., ed. *Physiological Basis of Aging and Geriatrics,* 2nd ed. Boca Raton, FL: CRC Press, 1994.

Timiras, P.S., W.B. Quay, and A. Vernakdakis, eds. *Hormones and Aging.* Boca Raton, FL: CRC Press, 1995.

Wild, C.P., and P. Pisani. "Carcinogen DNA and protein adducts as biomarkers of human exposure in environmental cancer epidemiology." *Cancer Detection and Prevention* 22 (1998): 273–283.

Willett, W.C., et al. "Intake of trans fatty acids and risk of coronary heart disease among women." *Lancet.* 341 (1993): 581–585.

Chapter 5. The Biology of Belief

Astin, A.W. "An integral approach to medicine." *Alternative Therapies in Health & Medicine* 8 (2002):70–75.

Austad, S.N., and K.E. Fischer. "Mammalian aging, metabolism, and ecology: evidence from the bats and marsupials." *Journal of Gerontology.* 46 (1991): B47–53.

Bale, T.L., F.J. Giordano, and W.W. Vale. "A new role for corticotropin-releasing factor-2. suppression of vascularization." *Trends in Cardiovascular Medicine* 13 (2003): 68–71.

Benson, H., et al. "Three case reports of the metabolic and electroencephalographic changes during advanced Buddhist meditation techniques." *Behavioral Medicine* 16 (1990): 90–95.

Blanc, S., et al. "Energy expenditure of rhesus monkeys subjected to 11 years of dietary restriction." *Journal of Clinical Endocrinology & Metabolism* 88 (2003): 16–23.

Black, S., et al. "Inhibition of Mantoux reaction by direct suggestion under hypnosis." *British Medical Journal* 1 (1963): 1649–1652.

Bordone, L., and L. Guarente. "Calorie restriction, SIRT1 and metabolism: understanding longevity." *Nature Reviews: Molecular and Cell Biology* 6 (2005): 298–305.

Bushell, W.C. "Evidence that a specific meditational regimen may induce adult neurogenesis." [abstract] *Developmental Brain Research* 132 (2001): A26.

Bushell, W.C., "From molecular biology to anti-aging cognitive-behavioral practices: the pioneering research of Walter Pierpaoli on the pineal and bone marrow foreshadows the contemporary revolution in stem cell and regenerative biology." *Annals of the New York Academy of Sciences* 1057 (Dec 2005): 28–49.

Bushell, W.C. Model: Potential cognitive-behavioral stem cell activation in multiple niches. Poster presented at the March 2005 Stem Cell Biology & Human Disease Conference, UCSD/Salk Institute/Nature Medicine, La Jolla, CA.

Bushell, W.C. "Psychophysiological and comparative analysis of ascetico-meditational discipline: toward a new theory of asceticism." *Asceticism.* V. L. Wimbush and R. Valantasis, eds. New York: Oxford University Press [Oxford Reference Series], 553–575.

Buxton, O.M., et al. "Acute and delayed effects of exercise on human melatonin secretion." *Journal of Biological Rhythms* 12 (1997): 568–574.

Cadet, P., et al. "Cyclic exercise induces anti-inflammatory signal molecule increases in the plasma of Parkinson's patients." *International Journal of Molecular Medicine* 12 (2003): 485–492.

Carlson, L.E., et al. "Mindfulness-based stress reduction in relation to quality of life, mood, symptoms of stress, and immune parameters in breast and prostate cancer outpatients." *Psychosomatic Medicine* 65 (2003): 571–581.

Carlson, L.E., et al. "Mindfulness-based stress reduction in relation to quality of life, mood, symptoms of stress and levels of cortisol, dehydroepiandrosterone sulfate (DHEAS) and melatonin in breast and prostate cancer outpatients." *Psychoneuroendocrinology* 29 (2004): 448–474.

Carrasco, G.A., and L.D. Van de Kar. "Neuroendocrine pharmacology of stress." *European Journal of Pharmacology* 2003;463:235–272

Carrillo-Vico, A., et al. "A review of the multiple actions of melatonin on the immune system." *Endocrine* 27 (2005): 189–200.

Chapman, L.F. "Changes in tissue vulnerability induced during hypnotic suggestion". *Journal of Psychosomatic Research* (1959) 499–505.

Coker, K.H. "Meditation and prostate cancer: integrating a mind/body intervention with traditional therapies." *Seminars in Urologic Oncology* 17 (1999): 111–118.

Davenport, J. *Animal Life at Low Temperature.* London: Chapman & Hall, 1992.

Epel, E.S., et al. "Accelerated telomere shortening in response to life stress." *Proceedings of the National Academy of Sciences* 101 (2004): 17312–17315.

Esch, T., et al. "Stress in cardiovascular disease." *Medical Science Monitor* 8 (2002): RA93–RA101.

Farrar, W.L., et al. "Visualization and characterization of interleukin-1 receptors in the brain." *Immunology* 139 (1987): 459–463.

Frolkis, V.V. "Stress-age syndrome." *Mechanisms of Ageing and Development* 69 (1993): 93–107.

Ganong, W.F. "The adrenal medulla and adrenal cortex." *Medical Physiology.* Norwalk, CT: Appleton & Lange, 1993, 334–335, 339–340.

Glaser, J.L., et al. "Elevated serum dehydroepiandrosterone sulfate levels in practitioners of the transcendental meditation (TM) and TM-Sidhi programs." *Journal of Behavioral Medicine* 15 (1992): 327–341.

Goto, S., et al. "Regular exercise: an effective means to reduce oxidative stress in old rats." *Annals of the New York Academy of Sciences* 1019 (2004): 471–474.

Greaves, G. "Reflections on a new medical cosmology." *Journal of Medical Ethics* 28 (2002): 81–85.

Gredilla, R., and G. Barja. "Minireview: the role of oxidative stress in relation to caloric restriction and longevity." *Endocrinology* 146 (2005): 3713–3717.

Hardeland, R. "Antioxidative protection by melatonin: multiplicity of mechanisms from radical detoxification to radical avoidance." *Endocrine* 27 (2005): 119–130.

Harinath, K., et al. "Effects of hatha yoga and Omkar meditation on

cardiorespiratory performance, psychologic profile, and melatonin secretion." *Journal of Alternative and Complementary Medicine* 10 (2004): 261–
268.

Heller, H.C., et al. "Voluntary hypometabolism in an Indian yogi." *Journal of
Thermal Biology* 12 (1987): 171–173.

Helm, H.M., et al. "Does private religious activity prolong survival? A six-
year follow-up of 3,851 older adults." *The Journals of Gerontology, Series A:
Biological Sciences and Medical Science* 55 (2000): M400–405.

Holloszy, J.O., and K.B. Schechtman. "Interaction between exercise and food
restriction: effects on longevity of male rats." *Journal of Applied Physiology.* 70 (1991): 1529–1535.

Holloszy, J.O. "Exercise increases average longevity of female rats despite
increased food intake and no growth retardation." *Journal of Gerontology*
48 (1993): B97–100.

Jacobs, G.D. "The physiology of mind-body interactions: the stress response
and the relaxation response." *Journal of Alternative and Complementary
Medicine* 8 (2002): 219.

Ji, L.L. "Exercise-induced modulation of antioxidant defense." *Annals of the
New York Academy of Sciences* 959 (2002): 82–92.

Jones, B.M. "Changes in cytokine production in healthy subjects practicing
Guolin Qigong: a pilot study." *BMC Complementary and Alternative Medicine.* 1 (2001): 8–12.

Kim, D.H., et al. "Effect of Zen meditation on serum nitric oxide activity and
lipid peroxidation." *Progress in Neuro-Psychopharmacology and Biological
Psychiatry* 29 (2005): 327–331.

Mason, A.A. "A case of congenital Ichthyosiform Erythrodermia of Brocq
treated by hypnosis." *British Medical Journal* 30 (1952): 442–443.

Sapolsky, Robert M., Lewis C. Krey, and Bruce S. McEwen. "The neuroen-
docrinology of stress and aging: The glucocorticoid cascade hypothesis."
Science of Aging Knowledge Environment (38) (September 25 2002) 21.

Shenefelt, P.D. "Biofeedback, cognitive-behavioral methods, and hypnosis in
dermatology: is it all in your mind?" *Dermatologic Therapy* 16:2 (2003):
114–22.

Tiller, W.A. *Science and Human Transformation: Subtle Energies, Intentionality and Consciousness.* Walnut Creek, CA: Pavior Publishing, 1997.

Tolle, Eckhart. *The Power of Now: A Guide to Spiritual Enlightenment.* Novato, CA: New World Library, 1999

Chapter 6. Personal Genetic Health Nutraceutical Supplement Use

Abbati, C., et al. "Nootropic therapy of cerebral aging." *Advances in Therapy;*8 (1991): 268–275.

Abraham, G.E., and J.D. Flechas. "Management of fibromyalgia: rationale for the use of magnesium and malic acid." *Journal of Nutritional Medicine* 3 (1992): 49–59.

Agren, J.J., et al. "Fish diet, fish oil and docosahexaenoic acid rich oil lower fasting and postpranadial plasma lipid levels." *European Journal of Clinical Nutrition* 50 (1996): 765–771.

Akesson, C., et al. "An extract of *Uncaria tomentosa* inhibiting cell division and NF-κB activity without inducing cell death." *International Immunopharmacology* 3 (2003):1889–1900.

Akesson, C., R.W. Pero, and F. Ivars. "C-Med-100, a hot water extract of *Uncaria tomentosa,* prolongs leukocyte survival *in vivo.*" *Phytomedicine* 10 (2003): 25–33.

Albert, C.M., et al. "Blood levels of long-chain fatty acids and the risk of sudden death." *New England Journal of Medicine* 346 (2002): 1113–1118.

Ames, B.A., et al. "High-dose vitamin therapy stimulates variant enzymes with decreased coenzyme binding affinity (increaded Km): relevance to genetic disease and polymorphisms." *American Journal of Clinical Nutrition* 75 (2002): 616–658.

Ames, B.A. "The Metabolic Tune-up: Metabolic Harmony and Disease Prevention." *Journal of Nutrition* 133 (2003) 1544S-1548S.

Ames, B.A. "Micronutrients prevent cancer and delay aging." *Toxicology Letters* 102–103 (Dec 1998): 5–18.

Ames, B.A., and P. Wakimoto. "Are vitamin and mineral deficiencies a major cancer risk?" *Nature Reviews: Cancer* 2 (2002): 694–704.

Ames, B.N., L. Elson-Schwab, and E.A. Silver. "High-dose vitamin stimulates variant enzymes with decreased coenzyme binding affinity (increased Km): relevance of genetic disease and polymorphisms." *American Journal of Clinical Nutrition* 75 (2002): 616–658.

Bagchi, M., et al. "Smokeless tobacco, oxidative stress, apoptosis and antioxidants in human oral keratinocytes." *Free Radical Biology and Medicine* 26 (1999): 992– 1000.

Barbiroli, B., et al. "Lipoic (thioctic) acid increases brain energy availability and skeletal muscle performance as shown in vivo 31P-MRS in a patient with mito-chondrial cytopathy." *Journal of Neurology* 2423 (1995): 472–477.

Barclay, L. "Dietary omega-3 fatty acids may reduce risk of age-related macular degeneration." [article] *Medscape Medical News,* May 14, 2003.

Barnes, P.J., and M. Karin. "Nuclear factor-KB—a pivotal transcription factor in chronic inflammatory diseases." *New England Journal of Medicine* 336 (1997): 1066–1071.

Barringer, T.A., et al. "Effect of a multivitamin and mineral supplement on infection and quality of life." *Annals of Internal Medicine* 138 (2003): 365–371.

Beecher, G.R. "Phytonutrients' role in metabolism: effects on resistance to degenerative processes." *Nutrition Reviews* 57 (1999): S3–S6.

Beg, A.A., and D. Baltimore. "An essential role for NF-kappa B in preventing TNF-alpha-induced cell death." *Science* 274 (1996): 782–784.

Bianchi, L., et al. "Carotenoids reduce the chromosol damage induced by bleomycin in human cultured lymphocytes." *Anticancer Research* 13 (1993): 1007–1010.

Bliznakov, E.G. "Coenzyme Q_{10}, lipid-lowering drugs (statins) and cholesterol: a present day Pandora's box." *JAMA, the Journal of the American Medical Association* 5 (2002): 32–38.

Bliznakov, E.G., and D.J. Wilkins. "Biochemical and clinical consequences of inhibiting coenzyme Q_{10} biosynthesis by lipid-lowering HMG-CoA reductase inhibitors (statins): a critical overview." *Advances in Therapy* 15 (1998): 218–228.

Bonnesen, C., I.M. Eggleston, and D. Hayes. Dietary indoles and isothio-cyanates that are generated from cruciferous vegetables can both stimulate apoptosis and confer protection against DNA damage in human colon cell lines. *Cancer Research* 61 (2001): 6120–6130.

Borek, C., "Antioxidant health effects of aged garlic extract." *Journal of Nutrition* 131 (2001): 1010S-1015S.

Borum, P.R. "Carnitine." *Annual Review of Nutrition* 3 (1983): 233–259.

Boyonoski, A.C., et al. "Niacin deficiency increases the sensitivity of rats to the short and long term effects of ethylnitrosourea treatment." *Molecular and Cellular Biochemistry* 193 (1999): 83–87.

Boyonoski, A.C., et al. "Pharmacological intakes of niacin increase bone marrow poly(ADP-ribose) and the latency of ethylnitrosourea-induced car-cinogenesis in rats." *Journal of Nutrition* 132 (2002): 115–120.

Bradlow, H.L., et al. "Multifunctional aspects of the action of indole-3–carbinole as an antitumor agent." *Annals of the New York Academy of Sciences* 889 (1999): 204–213.

Broquist, H.P. "Carnitine." *Modern Nutrition in Health and Disease.* Shils, ed., 8th ed. Philadelphia, PA: Lea & Febinger, 1994, 459–465.

Burge, B. "Mitochondria: the beleaguered powerhouse of the cell." *Health-line* (1999): 10–11.

Cao, L.Z., and Z.B. Lin. "Regulatory effect of *Ganoderma lucidum* polysac-charides on cyto toxic T-lymphocytes induced by dendritic cells in *vilor.*" *Acta Pharmacologica Sinica* 24 (2003): 312–326.

Ceda, G.P., et al. "Alpha-glycerylphosphorylcholine administration increases the GH responses to GHRH of young and elderly subjects." *Hormone and Metabolic Research* 24 (1991): 119–121.

Cerhan, J.R., et al. "Antioxidant micronutrients and risk of rheumatoid arthri-tis in a cohort of older women." *American Journal of Epidemiology* 157 (2003): 345–354.

Chambers, J.C., et al. "Improved vascular endothelial function after oral B vitamins: an effect mediated through reduced concentrations of free plas-ma homocysteine." *Circulation* 102 (Nov 2000): 2479–2483.

Chandra, R.K. "Effect of vitamin and trace-element supplementation on immune responses and infection in elderly subjects." *Lancet* 340 (1992): 1124–1127.

Chen, D.Z., et al. "Indole-3–carbinol and diindolylmethane induce apoptosis of human cervical cancer cells and in murine HPV16–transgenic preneoplastic cervical epithelium." *Journal of Nutrition* 131 (2001): 3294–3302.

Chew, B.P., et al. "A comparison of the anticancer activities of beta-carotene, canthaxanthin and astaxanthin in mice in vivo." *Anticancer Research* 19 (1999): 1849–1853.

Clarke, S.D., et al. "Fatty acid regulation of gene expression: its role in fuel partitioning and insulin resistance." *Annals of the New York Academy of Sciences* 827 (1997): 178–187.

Chopra, R.K., et al. "Relative bioavailability of coenzyme Q_{10} formulations in human subjects." *International Journal for Vitamin and Nutrition Research* 68 (1997): 109–113.

Giampapa, V., R.W. Pero, and M. Zimmerman. *The Anti-Aging Solution: 5 Simple Steps to Looking and Feeling Young.* Hoboken, NJ: Wiley, 2004.

Pero, R.W., V. Giampapa, and A. Vojdani. "Comparison of a broad spectrum anti-aging nutritional supplement with- and without the addition of a DNA repair enhancing Cat's Claw extract." *Journal of Anti-Aging Medicine* 5:4 (2002): 345– 353.

Chapter 7. The Full Personal Genetic Health Program

Atherden, S.M. "Development and application of a direct radioimmunoassay for aldosterone in saliva." *Steroids* 46 (1985): 845–855.

Banne, A.F., Amiri, A., and R.W. Pero. "Reduced level of serum thiols in patients with a diagnosis of active disease." *Journal of Anti-Aging Medicine* 6:4 (Dec 2003): 327–334.

Barrett-Connor, E. "Bioavailable testosterone and depressed mood in older men: the Rancho Bernardo Study." *Journal of Clinical Endocrinology & Metabolism* 84 (1999): 573–577.

Baruchel, S., and M.A. Wainberg. "The role of oxidative stress in disease progression in individuals infected by the human immunodeficiency virus." *Journal of Leucocyte Biology* 52 (1992): 111–114.

Beckman, K.B., and B.N. Ames. "The free radical theory of aging matures." *Physiological Reviews* 78 (1990): 547–581.

Blazejová, K., et al. "Sleep disorders and the 24–hour profile of melatonin and cortisol." *Sb Lek* (2000): 101347–101351.

Block, G., et al. "Factors associated with oxidative stress in human populations." *American Journal of Epidemiology* 156 (2002): 274–285.

Bohr, V., et al. "Oxidative DNA damage processing and changes with aging." *Toxicology Letters* 102–103 (1998): 47–52.

Bolaji, I.I. "Sero-salivary progesterone correlation." *International Journal of Gynecology* & *Obstetrics* 45 (1994): 125–131.

Brenner, D.D., et al. "Biomarkers in styrene-exposed boat builders." *Mutation Research* 261 (1991): 225–236.

Cedard, L., et al. "Progesterone and estradiol in saliva after in vitro fertilization and embryo transfer." *Fertility and Sterility* 47 (1987): 278–283.

Cortopassi, G.A., and E. Wang. "There is substantial agreement among interspecies estimates of DNA repair activity." *Mechanisms of Ageing and Development* 91:3 (1996): 211–218.

Dabbs, J.M., et al. "Reliability of salivary testosterone measurements: a multicenter evaluation." *Clinical Chemistry* 41 (1995): 1581.

Dabbs, J.M., et al. "Salivary testosterone measurements: reliability across hours, days, and weeks." *Physiology* & *Behavior* 48 (1990): 83–86.

Devaraj, S., et al. "Divergence between LDL oxidative susceptibility and urinary F2–isoprostanes as a measure of oxidative stress in type 2 diabetes." *Clinical Chemistry* 47 (2001): 1974–1978.

Dworski, R., et al. "Assessment of oxidant stress in allergic asthma by measurement of the major urinary metabolite of F2–isoprostane, 15–F2t-IsoP (8–iso PGF2alpha)." *Clinical* & *Experimental Allergy* 31 (2001): 387–390.

Giubilei, F., et al. "Altered circadian cortisol secretion in Alzheimer's disease: clinical and neuroradiological aspects." *Journal of Neuroscience Research* 66 (2001): 262–265.

Gotovtseva, L.P., and G.F. Korotko. "Salivary thyroid hormones in evaluation of the functional state of the hypophyseal-thyroid system." *Klinicheskaia laboratornaia diagnostika* 7 (2002): 9–11.

Grossi, G., et al. "Associations between financial strain and the diurnal salivary cortisol secretion of long-term unemployed individuals." *Integrative Physiological and Behavioural Science* 36 2001: 205–219.

Grube, K. and A. Burkle. "Poly(ADP-ribose) polymerase activity in mononuclear leukocytes of 13 mammalian species correlates with species-specific life span." *Proceedings of the National Academy of Sciences* 89 (1993): 11759–11763.

Ishikawa, M., et al. "The clinical usefulness of salivary progesterone measurement for the evaluation of the corpus luteum function." *Gynecologic and Obstetric Investigation* 53 (2002): 32–37.

Johnson, S.G., et al. "Direct assay for testosterone in saliva: relationship with a direct serum free testosterone assay." *Clinica Chimica Acta* 163 (1987): 309–318.

Klentrou, P., et al. "Effect of moderate exercise on salivary immunoglobulin A and infection risk in humans." *European Journal of Applied Physiology* 87 (2002): 153–158.

Laudat, M.H., et al. "Salivary cortisol measurement: a practical approach to assess pituitary-adrenal function." *Journal of Clinical Endocrinology & Metabolism* 66 (1988): 343–348.

Lawrence, H.P. "Salivary markers of systemic disease: noninvasive diagnosis of disease and monitoring of general health." *Journal of the Canadian Dental Association* 68 (2002): 170–174.

Lemaire, I., et al. "Stimulation of interleukin-1 and -6 production in alveolar macrophages by the neotropical liana, *Uncaria tomentosa* (Una de Gato)." *Journal of Ethnopharmacology* 64 (1999): 109–115.

Lewy, A.J. "The dim light melatonin onset, melatonin assays and biological rhythm researched in humans." *Biological Signals and Receptors* 8 (1999): 79–83.

Lieber, M.R. "Pathological and physiological double-strand breaks: roles in cancer, aging, and the immune system. [Warner-Lambert/Parke Davis Award Lecture] *American Journal of Pathology* 153 (1998): 1323–1332.

Lu, Y., et al. "Salivary estradiol and progesterone levels in conception and nonconception cycles in women: evaluation of a new assay for salivary estradiol." *Fertility and Sterility* 72 (1999): 951–952.

Mandel, I.D., et al. "The diagnostic uses of saliva." *Journal of Oral Pathology & Medicine* 19 (1990): 119–125.

Mayer, J., et al. "Biologic markers in ethylene oxide, exposed workers and controls." *Mutation Research* 248 (1991): 163–176.

Mejtek, V.A. "High and low emotion events influence emotional stress perceptions and are associated with salivary cortisol response change in a consecutive stress paradigm." *Psychoneuroendocrinology* 27 (2002): 337–352.

Miletic, I.D., et al. "Salivary IgA secretion rate in young and elderly persons." *Physiology & Behavior* 60 (1996): 243–248.

Morgan, C.A., et al. "Hormone profiles in humans experiencing military survival training." *Biological Psychiatry* 47 (2000): 891–901.

Nagetgaal, E., et al. "Correlation between concentrations of melatonin in saliva and serum in patients with delayed sleep phase syndrome." *Therapeutic Drug Monitoring* 20 (1998): 181–183.

Obminski, Z., and R. Stupnicki. "Comparison of the testosterone-to-cortisol ratio values obtained from hormonal assays in saliva and serum." *Journal of Sports Medicine and Physical Fitness* 37 (1997): 50–55.

Obminski, Z., et al. "Effect of acceleration stress on salivary cortisol and plasma cortisol and testosterone levels in cadet pilots." *Journal of physiology and pharmacology* 48 (1997): 193–200.

Park, S.J., and H. Tokura. "Bright light exposure during the daytime affects circadian rhythms of urinary melatonin and salivary immunoglobulin A." *Chronobiology International* 16 (1999): 359–371.

Pearson, T.A., et al. "Markers of inflammation and cardiovascular disease: application to clinical and public health practice: a statement for healthcare professionals from the Centers for Disease Control and Prevention and the American Heart Association." *Circulation* 107 (2003): 499–511.

Pero, R.W., and V. Giampapa. "Oxidative stress and its effects on immunity and apoptosis—DNA repair as a primary molecular target for antiaging therapies." (2001 position paper, unpublished).

Index